Orthogonal Design in Concrete Application

蔡正詠　王足献　著

王贤忠　梅占敏　孟繁杰　郭春雷　译
张　珏　王　伟　赵彦贤

黄河水利出版社
·郑州·

Summary

Based on the experimental study of cement concrete, this book systematically introduces the basic method of orthogonal design and its application in concrete from the point of view of mathematical statistics, mainly including optimizing the mineral composition of high-strength Portland cement clinker, optimizing the variety and quantity of water-reducing agent, configuring special concrete and selecting concrete mix. The application of mixing ratio, researching concrete performance, optimizing concrete forming, steam curing technology and so on.

This book can be used for reference by researchers of concrete materials and construction technicians of concrete works.

图书在版编目(CIP)数据

正交设计在混凝土中的应用 = Orthogonal Design in Concrete Application：英文/蔡正詠，王足献著；王贤忠等译. —郑州：黄河水利出版社，2018.11

ISBN 978－7－5509－2217－4

Ⅰ.①正… Ⅱ.①蔡… ②王… ③王… Ⅲ.①正交设计－应用－混凝土工程－英文 Ⅳ.①TU755

中国版本图书馆 CIP 数据核字(2018)第 281730 号

组稿编辑：简群　　电话：0371-66026749　　E-mail:931945687@qq.com

出　版　社：黄河水利出版社　　　　　　　　　　　网址：www.yrcp.com
　　　　　　地址：河南省郑州市顺河路黄委会综合楼 14 层　邮政编码：450003
发行单位：黄河水利出版社
　　　　　　发行部电话：0371-66026940、66020550、66028024、66022620(传真)
　　　　　　E-mail:hhslcbs@126.com
承印单位：河南新华印刷集团有限公司
开本：787 mm×1 092 mm　1/16
印张：11.75
字数：360 千字　　　　　　　　　　　　　印数：1—1 000
版次：2018 年 11 月第 1 版　　　　　　　　印次：2018 年 11 月第 1 次印刷

定价：65.00 元

Preface

With the development of socialist construction in our country, the demand and usage of concrete materials is increasing day by day. Especially new raw materials and new workers are emerging constantly, and more and more concrete materials are being tested. The traditional concrete material test is characterized by: factors: long multi-cycle, large error fluctuations, heavy workloads, test data often messy, difficult to analyze. Practice shows that the concubine is well tested. No need to do many tests, you can get satisfactory results. If poorly arranged, more times and results are not necessarily satisfactory. The number of tests is unreasonably large, and a great deal of manpower and material resources will inevitably be wasted. Sometimes it may lead to conclusions that do not necessarily result from prolonged periods of time and changes in test conditions.

For a long time, the test of concrete materials mainly deal with the test data passively, and almost did not make any demands on the test arrangement. This not only resulted in the blind increase of the number of tests, but also the test often failed to provide sufficient and reliable information, Many multivariate trials fail to achieve their intended purpose. As some experimenters said: "This is a big screen to catch small fish and sometimes catch no nose." So how can we combine the experimental design with the data analysis so that we can do both less testing and more enrichment? The experimental information and can draw a comprehensive conclusion, is a problem worth studying. Applying orthogonal design to arrange concrete test is a good and effective way to research and solve this problem.

Orthogonal design has been widely used in many industries. Last few years come, in the concrete material test application, has achieved good results. For example, overseas countries have adopted orthogonal design in RCC, expansive agent, steam curing process and concrete performance test in hot area.

Orthogonal design has also been applied in concrete experiment in our country, and gratifying results have been achieved. For example, the recipe of CS and CAS coagulant and the technology parameters of the new technology of promoting condensate pressure steaming, the production process parameters of the high efficiency water reducer such as FDN are all found by orthogonal design. In another example, fly ash quality standards in the pilot study, has made nearly 300 groups of tests, the test data are numerous, but the results of a certain regularity, and later with orthogonal design, only used 24 of the test results Find out the various primary and secondary factors and a number of factors.

Although the application of orthogonal design in concrete materials testing has made preliminary quit, but the depth and breadth of application far can not meet the requirements. Therefore, it is very necessary to exchange experience in this area and vigorously publicize the relevant basic knowledge. In recent years, the author has been invited to participate in several lectures organized by the Society of Civil Engineering and the Institute of Silicate and related units, systematic introduction of this knowledge, a welcome book is in this case based on the majority of concrete materials test workers need to write according to the experience of some people who apply orthogonal design in concrete material test, we refer to the relevant materials and learn from the experience both at home and abroad. We have basically sorted out and selected about 30 representative examples, hoping that these examples can serve as a by-pass, The effect, so that readers can combine professional quickly grasp and use this method.

This book mainly according to professional needs prepared. On the one hand, consider various aspects of concrete materials professional, including from the research and application of raw materials to test the properties of specialty concrete, a total of ten aspects (see book appendix). On the other hand, but also on a case-by-case basis, from mathematical statistics Angle, the system introduces the basic methods of orthogonal design and the main experience of the application of orthogonal design. This book focuses on the application of the method, not for the strict mathematical derivation, readers who wish to understand the principles of mathematics; you can refer to the reference book attached to the book.

The book is divided into seven chapters and three appendices. Beginners to read the first two chapters to master the basic method, and then read the relevant sections if necessary, Appendix I lists ten representative examples for professional workers reference; Appendix II lists the book instance index, According to professional or orthogonal table types of different situations, identify the required reference examples; Appendix III lists the commonly used orthogonal tables and related statistical tables for readers to check.

For orthogonal design and related mathematical principles, we are still continuing to learn, because of the limited level, the book shortcomings and errors are inevitable, please reader criticism.

Some of the book's data is drawn from the research results of some units and individuals, to express my gratitude.

Author
October 1983

Contents

Chapter 1　Orthogonal Design Preliminary ⋯⋯⋯⋯⋯⋯⋯⋯⋯⋯⋯⋯⋯⋯⋯⋯⋯⋯⋯ 1
　1.1　Why Test Orthogonal Design ⋯⋯⋯⋯⋯⋯⋯⋯⋯⋯⋯⋯⋯⋯⋯⋯⋯⋯⋯⋯⋯⋯⋯⋯ 1
　1.2　Orthogonal Design of the Basic Method ⋯⋯⋯⋯⋯⋯⋯⋯⋯⋯⋯⋯⋯⋯⋯⋯⋯⋯⋯ 3
　1.3　The Basic Principles and Characteristics of Orthogonal Design ⋯⋯⋯⋯⋯⋯⋯⋯ 11
Chapter 2　Flexible Use Orthogonal Design ⋯⋯⋯⋯⋯⋯⋯⋯⋯⋯⋯⋯⋯⋯⋯⋯⋯⋯⋯ 17
　2.1　Orthogonal Design with Different Number of Levels ⋯⋯⋯⋯⋯⋯⋯⋯⋯⋯⋯⋯⋯ 17
　2.2　The Proposed Level of Law ⋯⋯⋯⋯⋯⋯⋯⋯⋯⋯⋯⋯⋯⋯⋯⋯⋯⋯⋯⋯⋯⋯⋯⋯ 20
　2.3　Activity Level and its Application ⋯⋯⋯⋯⋯⋯⋯⋯⋯⋯⋯⋯⋯⋯⋯⋯⋯⋯⋯⋯⋯ 23
　2.4　Composite Factors and Their Applications ⋯⋯⋯⋯⋯⋯⋯⋯⋯⋯⋯⋯⋯⋯⋯⋯⋯ 26
　2.5　Multi-indicator Orthogonal Design Analysis ⋯⋯⋯⋯⋯⋯⋯⋯⋯⋯⋯⋯⋯⋯⋯⋯⋯ 28
Chapter 3　Variance Analysis of Orthogonal Design ⋯⋯⋯⋯⋯⋯⋯⋯⋯⋯⋯⋯⋯⋯⋯ 35
　3.1　Introduction of Variance Analysis ⋯⋯⋯⋯⋯⋯⋯⋯⋯⋯⋯⋯⋯⋯⋯⋯⋯⋯⋯⋯⋯ 35
　3.2　Variance Analysis of Orthogonal Design ⋯⋯⋯⋯⋯⋯⋯⋯⋯⋯⋯⋯⋯⋯⋯⋯⋯⋯ 43
　3.3　Error Handling ⋯⋯⋯⋯⋯⋯⋯⋯⋯⋯⋯⋯⋯⋯⋯⋯⋯⋯⋯⋯⋯⋯⋯⋯⋯⋯⋯⋯⋯ 60
Chapter 4　Simple Variance Analysis of Orthogonal Design and Multiple Comparisons
⋯⋯⋯⋯⋯⋯⋯⋯⋯⋯⋯⋯⋯⋯⋯⋯⋯⋯⋯⋯⋯⋯⋯⋯⋯⋯⋯⋯⋯⋯⋯⋯⋯⋯⋯⋯⋯⋯⋯ 64
　4.1　Extreme and Mean Squares ⋯⋯⋯⋯⋯⋯⋯⋯⋯⋯⋯⋯⋯⋯⋯⋯⋯⋯⋯⋯⋯⋯⋯⋯ 64
　4.2　Application of Range Method in Variance Analysis ⋯⋯⋯⋯⋯⋯⋯⋯⋯⋯⋯⋯⋯ 66
　4.3　Multiple Comparison of T Method ⋯⋯⋯⋯⋯⋯⋯⋯⋯⋯⋯⋯⋯⋯⋯⋯⋯⋯⋯⋯ 71
Chapter 5　The Orthogonal Design of Interaction Function ⋯⋯⋯⋯⋯⋯⋯⋯⋯⋯⋯⋯ 76
　5.1　Interaction Concepts and Judgments ⋯⋯⋯⋯⋯⋯⋯⋯⋯⋯⋯⋯⋯⋯⋯⋯⋯⋯⋯⋯ 76
　5.2　Differences between Interaction and Experimental Error ⋯⋯⋯⋯⋯⋯⋯⋯⋯⋯⋯ 79
　5.3　The Experimental Arrangements and Analytical Methods with Interaction Function ⋯ 82
　5.4　Mixed Skills ⋯⋯⋯⋯⋯⋯⋯⋯⋯⋯⋯⋯⋯⋯⋯⋯⋯⋯⋯⋯⋯⋯⋯⋯⋯⋯⋯⋯⋯⋯⋯ 89
Chapter 6　Regression Analysis of Orthogonal Design ⋯⋯⋯⋯⋯⋯⋯⋯⋯⋯⋯⋯⋯⋯ 91
　6.1　Introduction of Regression Analysis ⋯⋯⋯⋯⋯⋯⋯⋯⋯⋯⋯⋯⋯⋯⋯⋯⋯⋯⋯⋯ 91
　6.2　Binary Orthogonal Regression Analysis ⋯⋯⋯⋯⋯⋯⋯⋯⋯⋯⋯⋯⋯⋯⋯⋯⋯⋯ 98
　6.3　Multivariate Orthogonal Regression Analysis ⋯⋯⋯⋯⋯⋯⋯⋯⋯⋯⋯⋯⋯⋯⋯⋯ 102
Chapter 7　Data Structure and Effect Estimation of Orthogonal Design ⋯⋯⋯⋯⋯⋯ 108
　7.1　Basic Concepts of Data Structure ⋯⋯⋯⋯⋯⋯⋯⋯⋯⋯⋯⋯⋯⋯⋯⋯⋯⋯⋯⋯⋯ 108
　7.2　Data Structure Orthogonal Design ⋯⋯⋯⋯⋯⋯⋯⋯⋯⋯⋯⋯⋯⋯⋯⋯⋯⋯⋯⋯⋯ 110
　7.3　Orthogonal Designs ⋯⋯⋯⋯⋯⋯⋯⋯⋯⋯⋯⋯⋯⋯⋯⋯⋯⋯⋯⋯⋯⋯⋯⋯⋯⋯⋯ 113
Appendix I　Supplementary Examples ⋯⋯⋯⋯⋯⋯⋯⋯⋯⋯⋯⋯⋯⋯⋯⋯⋯⋯⋯⋯⋯ 116
Appendix II　Whole Book Case Index ⋯⋯⋯⋯⋯⋯⋯⋯⋯⋯⋯⋯⋯⋯⋯⋯⋯⋯⋯⋯⋯ 148
Appendix III　Annexed Table ⋯⋯⋯⋯⋯⋯⋯⋯⋯⋯⋯⋯⋯⋯⋯⋯⋯⋯⋯⋯⋯⋯⋯⋯⋯ 153
Reference ⋯⋯⋯⋯⋯⋯⋯⋯⋯⋯⋯⋯⋯⋯⋯⋯⋯⋯⋯⋯⋯⋯⋯⋯⋯⋯⋯⋯⋯⋯⋯⋯⋯⋯ 182

Chapter 1
Orthogonal Design Preliminary

1.1 Why Test Orthogonal Design

Any concrete test to be done, there is how to arrange the test program and how to analyze the test results.

Look at a simple example.

【Example 1.1】 A unit in the development of alum stone water strength, taking into account the alunite (A), Anhydrous gypsum (B), and slag (C) are the main factors influencing the strength of alunite cement. It is intended to test each of these factors at three different levels (called levels) to determine the optimal formulation of the Ming Vanadium High Strength Cement. The factors and levels in the trials are listed in Table 1.1.1.

Table 1.1.1 Factor level table

Level	Factor		
	A. Alunite admixture (%)	B. Anhydrite content(%)	C. Slag content
1	8	5	13
2	10	7	11
3	12	9	6

For ease of description, the three levels of factor A are represented by A_1, A_2 and A_3, respectively; the same applies to B and C. Let us first recall how the experiment was conducted in the past. In the past, the following methods have been used in many cases:

First fixed B is B_1, C is C_1, change A, that is

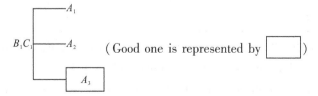

Three experiments found that A_3 is better. Then fix A for A_3, C for C_1, change B, that is

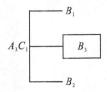

B_2 is better in the test results. Finally fixed A is A_3, B is B_2, change C, that is

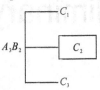

The result is C_2 is better. So concluded that $A_3B_2C_3$ is the best formula. This method of arranging trials is called the "isolated variable method." This method to arrange the test can also get some results, but its biggest drawback is the poor representation of the test. If these test points are plotted in Figure 1.3.2, you will see that the test points are located in a corner of the test range and there is no test point in the other test points. Therefore, it is not natural to arrange the test in this way, If there is an interaction between the factors that make up the formula, the method of arranging the experiment may not get the correct conclusion. Therefore, in the case of an interaction with all the combination of laurite, anhydrous gypsum and slag test, we call it a comprehensive test.

This example is a three-factor level test, if the three factors in Table 1.1.1. Of each level are met, you need to make $3^3 = 27$ tests, all 27 tests done, known as the "comprehensive test method", all the conditions of 27 tests are shown in Table 1.1.2.

Table 1.1.2 $3^3 = 27$ Combination conditions of experimental group

B	C	A		
		A_1	A_2	A_3
B_1	C_1	$A_1\ B_1\ C_1$	$A_2\ B_1\ C_1$	$A_3\ B_1\ C_1$
	C_2	$A_1\ B_1\ C_2$	$A_2\ B_1\ C_2$	$A_3\ B_1\ C_2$
	C_3	$A_1\ B_1\ C_3$	$A_2\ B_1\ C_3$	$A_3\ B_1\ C_3$
B_2	C_1	$A_1\ B_2\ C_1$	$A_2\ B_2\ C_1$	$A_3\ B_2\ C_1$
	C_2	$A_1\ B_2\ C_2$	$A_2\ B_2\ C_2$	$A_3\ B_2\ C_2$
	C_3	$A_1\ B_2\ C_3$	$A_2\ B_2\ C_3$	$A_3\ B_2\ C_3$
B_3	C_1	$A_1\ B_3\ C_1$	$A_2\ B_3\ C_1$	$A_3\ B_3\ C_1$
	C_2	$A_1\ B_3\ C_2$	$A_2\ B_3\ C_2$	$A_3\ B_3\ C_2$
	C_3	$A_1\ B_3\ C_3$	$A_2\ B_3\ C_3$	$A_3\ B_3\ C_3$

"Comprehensive test method" for things within the regularity can be more clearly analyzed, but the biggest drawback is the number of tests required too much can be seen from Table 1.1.3, the number of factors used in the test for each additional one, even if the two In the case of a horizontal one, the number of tests should be doubled, that is, when the number of tests increases by an equal number of times, the number of trials conducted in full will be more or less the same. In fact, it is often impossible to do so. In addition, the "isolated variable method" is used to arrange the test, which, if not repeated, gives no estimate of the experimental error.

Table 1.1.3 The number of test times

Factor	Level	Level number			
		2	3	4	5
Factor number	1	2	3	4	54
	2	4	9	16	25
	3	8	27	64	125
	4	16	81	256	625
	5	32	213	1 024	3 125
	6	64	729	4 096	15 625
	7	128	21 87	16 384	78 125
	8	256	6 561	65 536	390 625
	9	512	19 683	262 144	19 533 225
	10	1 024	59 049	1 048 576	9 765 625

From the comparison of the "isolated variable method" to the "comprehensive test method", we naturally ask: Is there any such method that can not only reduce the number of trials but also overcome the above shortcomings? The answer is yes. Orthogonal design is a good way to scientifically arrange a multi-factor test protocol and effectively separate test results. It absorbs the advantages of both methods and overcomes their shortcomings. Orthogonal design is the use of two sets of good orthogonal table, from a large number of multi-factorial comprehensive test selected less, but the Zen representative of the general conditions for the trial; test, taste less Test, and a simple calculation, you can find a better process conditions or optimal formula. In the next section, we will combine examples to introduce in detail the basic methods of applying orthogonal design to the test plan and analyzing the test results.

1.2 Orthogonal Design of the Basic Method

1.2.1 What is the Orthogonal Table?

Orthogonal table is the use of "balanced dispersion" and "neat comparable" two orthogonally principle, from a large number of test points to pick out a suitable test checkpoint, made of regularly arranged forms, this table becomes an orthogonal table. It is the basic tool for orthogonal design. We first recognize the following simple orthogonal tables.

$L_9(3^4)$ orthogonal table has 9 rows, 4 columns. The table consists of words "1", "2", "3". It has two characteristics: (1) Each column "1", "2", "3" appear the same number of times, all three times; (2) any two vertical columns, (1, 2), (1,3), (2,1), (2,2), (2,3), (3,1),

(3,2), (3,3) the same number of occurrences, are once or any two vertical alignment of the word "1", "2", "3" with a balanced.

$L_9(3^4)$

Experimental number	Column number			
	1	2	3	4
1	1	1	1	1
2	1	2	2	2
3	1	3	3	3
4	2	1	2	3
5	2	2	3	1
6	2	3	1	2
7	3	1	3	2
8	3	2	1	3
9	3	3	2	1

$L_4(2^3)$

Experimental number	Column number		
	1	2	3
1	1	1	1
2	1	2	2
3	2	1	2
4	2	2	1

$L_8(4^3 \times 2^4)$

Experimental number	Column number				
	1	2	3	4	5
1	1	1	2	2	1
2	3	2	2	1	1
3	2	2	2	2	2
4	4	1	2	2	2
5	1	2	1	1	2
6	3	1	1	2	2
7	2	1	1	1	1
8	4	2	1	2	1

$L_4(2^3)$ is the smallest orthogonal table, it also has two characteristics similar to $L_9(3^4)$: (1) There are two "1" and two "2" for each column; (2) Any two vertical (1, 1), (1, 2), (2, 1), (2, 2) of the four ordinal pairs formed in the horizontal direction appear once, that is, their collocations are also balanced. It has four rows, three columns.

$L_8(4^1 \times 2^4)$ is an orthogonal table with mixed levels (i.e. unequal levels). There are eight rows, five columns. The first column consists of the yards "1", "2", "3" and "4", while the remaining four columns are composed of the letters "1" and "2". The table has a factor of four to look at, other factors are two levels. There are still two characteristics similar to the above: (1) In each column, the number of occurrences is the same for each of the respective yards. (2) Any two columns with eight numbers Right, in terms of their respective yards, the same number of occurrences, that is, for any two vertical alignment between their words are also balanced. These two characteristics are the embodiment of the orthogonality principle of "balanced dispersion" and "neat comparability", that is, the meaning of orthogonality of orthogonal tables.

The meaning expressed by the orthogonal symbol is, for example, the following illustration. Common orthogonal tables see Appendix 1

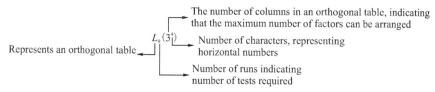

1.2.2 How to Design a Test Plan

Look at an example.

【Example 1.2.1】 In the trial production of strong concrete, the composite mixed with slag, gypsum and iron powder in order to enhance the degree of suspicion of soil. If you take three because of three levels, asked to choose the optimal dosage of each factor.

As mentioned above, if all the combinations at each level of each factor are tested, $3^3 = 27$ times, too many tests, can we make some tests and get good results? In other words, how to design the test program? The basic method of arranging the test protocol with orthogonal design is as follows:

1.2.2.1 Clear Test Purpose, to Determine Assessment Indicators

Before the test, first of all, we must first clear through the test to solve any problem, find out what the law. Following the p to determine the assessment indicators. The purpose of the test in Example 1.2.1 is to select the optimum dosage for each factor by experimenting with the law of the effect of slag, gypsum and iron powder on the concrete strength. In this way, the exam indicator is the compressive strength.

It must be noted that in the concrete test, we determine the assessment indicators, should be able to directly indicate the amount, such as strength, slump, weight loss rate, which are called quantitative indicators. Can not be used to express non-quantitative indicators, to find ways to become the amount of indicators (such as ratings, etc.) to be orthogonal design.

1.2.2.2 Select Factors, Select the Level, the Specified Factor Level Table

The so-called factor refers to the factors that affect the assessment index, the so-called level refers to the specific conditions of the factors to be compared in the experiment. Factors and levels are based

Orthogonal Design in Concrete Application

on test conditions, test purposes, and on a trial basis or by virtue of professional knowledge to determine. This example picks three factors, each of the three levels of selection, see Table 1.2.1. Table 1.2.1 factor A (slag love) three levels are: the level of 1 to 10%, the level of 2 to 15%, the level of 3 to 20%. Factors B (gypsum content) and factor C (iron powder content) of the three horizontal categories.

Table 1.2.1 Factor level table

Level	Factor		
	A. Slag content (%)	B. Anhydrite content (%)	C. Iron powder (%)
1	10	2	3
2	15	3.5	6
3	20	5	9

Note: The amount of cement accounted for the amount of %.

Among the various factors, the level can be expressed as the amount of use, known as quantitative factors; can not be used to represent, such as cement varieties, super plasticizer species, called qualitative factors. The level of each factor can be equal or not equal. Important factors or needs to focus on the investigation can choose some level; the rest may be less choice. For each cable, which corresponds to which level with firefly, it can be specified, in general, it is best to disrupt the order to arrange.

It is important to note that the factors are constrained by a relation in cement-concrete mix testing. Only selected (all factors-1) factors to investigate. For example, the laurite cement ratio test, all four factors, namely, clinker + slag + gypsum + iron dressing = 100% to form a relationship, only selected slag, gypsum, iron powder three factors, clinker this The factor is not independent, otherwise it can not satisfy this Wu system. Another example is the concrete mix test, a group of material composition is subject to a relationship between the constraints that the sum of cement, water, sand and stone content equal to the concrete bulk density ($U = 2\ 400\ kg/m^3$) or the absolute volume of all materials and equal to 1 000 L. Only choose one of the water, water-cement ratio and the amount of stones, the amount of sand this factor is not independent, otherwise it can not meet this relationship.

1.2.2.3 Make Use of Orthogonal Table to Arrange the Experiment

According to the level of factors table, the specific steps to arrange the test program are as follows:
1. Select the orthogonal table of the counterparts. When selecting the table, note that the number of levels in the factor level table should be exactly the same as the number of levels in the counterpart orthogonal table. However, the number of factors in the factor level table can be less than or equal to the number of columns in the counterpart orthogonal table. This example is a three-factor, threelevel test, the right female counterpart is $L_9(3^4)$. Use this table for 9 tests.
2. Fill in the specific factors to the selected orthogonal table.
   a. The order of the factors above: according to the factor level table fixed order, the A, B, C various factors in order to $L_9(3^4)$ on the first two columns 1, 2, 3, the fourth column is empty, with To estimate the experimental error.

Chapter 1 Orthogonal Design Preliminary

b. Horizontal Condemnation: After the factors are fixed on the columns, factors are respectively put behind the words "1", "2" and "3" of the three columns where the factors A, B and C are located the specific level of the horizontal table, here to condemnation.

For example, in column 1, the back of "1" is filled with 10%, the back of "2" is filled with 15% and the back of "3" is filled with 20%. Column 2, column 3 is similar to column 1. Thus Table 1.2.2 becomes Example 1.2.1: Test Plan.

3. List the test conditions Table 1.2.2 9 rows of concrete is the specific conditions of 9 tests. For example, No.1 test conditions are: slag content of 10%, gypsum content of 2%, and iron powder content of 3%. The remaining eight test number according to the above method to determine the test conditions.

After the test plan is confirmed, the test is carried out strictly according to the test conditions. It should be pointed out that: In addition to the selected due to the cold, the other conditions should be fixed in order to make a reasonable comparison.

Table 1.2.2 $L_9(3^4)$ Test result of extreme difference calculation

Experimental number	A. Slag content 1	B. Anhydrite content 2	C. Iron powder 3	4	28-day Compressive strength (kg/cm²)
1	1(10%)	1(2%)	1(3%)	1	765
2	1(10%)	2(3.5%)	2(6%)	2	810
3	1(10%)	3(5%)	3(9%)	3	858
4	2(15%)	1(2%)	2(6%)	3	857
5	2(15%)	2(3.5%)	3(9%)	1	891
6	2(15%)	3(5%)	1(3%)	2	765
7	3(20%)	1(2%)	3(9%)	2	907
8	3(20%)	2(3.5%)	1(3%)	3	867
9	3(20%)	3(5%)	2(1%)	1	860
K_1	2 333	2 529	2 397	2 516	
K_2	2 513	2 568	2 527	2 482	SUM: 7 480
K_3	2 634	2 383	2 566	2 482	
$\overline{K}_1$	778	843	799	838	
$\overline{K}_2$	838	856	842	827	
$\overline{K}_3$	878	794	852	827	
R	100	62	53	12	

(Tsinghua University, 1977)

1.2.3 How to Analyze the Test Results

The compressive strength results of 9 tests are recorded on the right of Table 1.2.2.

1. Look directly.

As you can see from Table 1.2.2, the strength of Test No. 7 is 907 kg/cm². The test conditions are: $A_3B_1C_3$, such as slag content of 20%, gypsum content of 2%; iron dressing content of 9%.

2. Calculate.

Through simple calculation can reach the following (a) analysis of the relationship between strength and strength, that is, when the level of factors change, how to change the strong virtual; (b) analysis of factors affecting the strength of the order of the primary, that is, in the three factors Which of the seven major factors that affect the productivity, which is a secondary factor (c) to determine the optimal blending content, that is, what level of the three factors, the concrete strength is the highest; (d) pointed out for further tests; (e) If available, you can estimate the size of the test error.

Under each column of Table 1.2.2, calculate the three corresponding levels: the sum of test intensities K_1, K_2, K_3 and the average intensity, $\overline{K}_1$, $\overline{K}_2$, $\overline{K}_3$ and its range R, and its calculation method as follows:

For the first column K_i and $\overline{K}_i$ values

$K_1 = 765 + 810 + 758 = 2\ 333$ (No.1, No.2, No.3 experiment strength)

$K_2 = 857 + 891 + 765 = 2\ 513$ (No.4, No.5, No.6 experiment strength)

$K_3 = 907 + 867 + 860 = 2\ 634$ (No.7, No.8, No.9 experiment strength)

$$\overline{K}_1 = \frac{K_1}{3} = \frac{2\ 333}{3} = 778$$

$$\overline{K}_2 = \frac{K_2}{3} = \frac{2\ 513}{3} = 838$$

$$\overline{K}_3 = \frac{K_3}{3} = \frac{2\ 634}{3} = 878$$

The calculation of $K_i(K)$ values for other columns (including null columns) is the same as for column 1.

The calculation of the K_i value is correct and can be calculated using the following equation:

$K_1 + K_2 + K_3 = 7\ 480$ (total of 9 test intensities)

If not established, that is to find out the error.

The range of each column R, from each column $\overline{K}_1$, $\overline{K}_2$, $\overline{K}_3$ (or K_1, K_2, K_3) three with the number of the largest reduction. For example, the first column $R = 878 - 788 = 100$, the other columns are calculated the same way as the first column.

After drawing the trend chart calculation range is poor, the quantitative level of the three levels of hope factors, the general should draw the dose and the average intensity (or intensity and) of the relationship between the graph directly to see the strength changes with the amount of each factor General relationship. The actual amount of each factor (rather than the size of the horizontal number) as the horizontal axis, the average intensity of the vertical axis, draws the trend of various factors (see Figure 1.2.1).

Chapter 1 Orthogonal Design Preliminary

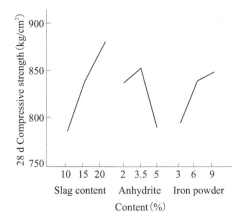

Figure 1.2.1 *uses of each factor and the relationship between strength*

After making a simple calculation, how to carry on the next analysis?

For each column, comparing the magnitudes of their average intensities $\overline{K}_1$, $\overline{K}_2$, $\overline{K}_3$, if they are larger than $\overline{K}_1$, $\overline{K}_2$, $\overline{K}_3$, occupies the level 3 of the set of factors, which is generally better than levels 1 and 2 in intensity. As in the first column, $\overline{K}_3 = 878$, which is larger than both $\overline{K}_1 = 778$ and $\overline{K}_2 = 838$, this roughly indicates that 20% of the slag content is better than 10% and 15% of the slag content.

The size of the range of very poor R, used to measure the size of the test due to. Very large factors, indicating that the strength of its three levels caused by the large differences, usually an important factor, and very small factors, is often a secondary factor. Eucalyptus according to the size of extremely poor, in this case the order of the primary and secondary factors: $A \to B \to C$, that is, the amount of is the main factor affecting the strength of stone content is a minor factor, while the amount of iron powder Smaller.

This relationship can also be reflected in the trend chart, with the largest fluctuation of the A-factor graph, which is the main factor; the B-factor graph fluctuates less and is the secondary factor and the C-factor graph fluctuates the smallest, which is the third influencing factor.

The empty range of the poor $R = 12$ kg/cm^2, as the test error estimate. The empty column in this case is very small, indicating that the accuracy of the test is quite high. If the null column is more than two columns, for the equal level of the orthogonal table, the average of the empty column with no interaction between the factors is used as the estimation of the experimental error. The reason why the difference of null columns can be used as an estimation of experimental error is discussed in Chapter Three.

According to the range analysis (or visual analysis) may come to a good combination of conditions are $A_3B_2C_3$.

3. The best dosage (or formula) to determine

$A_3B_1C_3$ is a good condition to be directly seen, however, the good condition for the calculation is $A_3B_2C_3$, but it is not included in the 9 tests. In order to demonstrate the conclusions drawn from the analysis, it is generally advisable to select the likely combination condition $A_3B_2C_3$ in parallel with the combination condition of the original No. 7 test, on the one hand, and on the other hand for their reproducibility and reliability. However, in this case, as we have already analyzed, the factor is the secondary factor. The difference in strength between $\overline{K}_2$ and $\overline{K}_1$ is not big, about 13 kg/cm^2. In order to save the raw materials and reduce the cost, we should choose the best one the amount of $A_3 B_1 C_3$, is the No. 7 test conditions.

With intuitive analysis of orthogonal designs, the set of basic methods that optimize the recipe or process parameters ends here.

In the concrete experiment, the application of orthogonal design in addition to solve similar Example 1.2.1 Yong problems, but also use it to focus on examination because of the intrinsic norms with the evaluation index to seek better production conditions as a reasonable guide Practice put forward the direction. Let's look at an example below.

【Example 1.2.2】 examining the impact of cement varieties, initial squareness of concrete and temperature at the point of agitation on slump loss[7].

Assessment indicators: Slump loss as small as possible.

1. Factors and levels are shown in Table 1.2.3.

Table 1.2.3 Factor level table

Level	Factor		
	A. Cement varieties	B. Slump(cm)	C. Outlet temperature(℃)
1	A	High(18 – 19)	21
2	B	Low(8 – 9)	29

Note: $A - C_3 A + C_3 S + C_2 S + C_4 AF + C_4 AF = 10.6 + 52.5 + 17.5 + 8.0(\%)$
$B - C_3 A + C_3 S + C_2 S + C_4 AF + C_4 AF = 8.9 + 54 + 21.1 + 8.2(\%)$

2. Orthogonal table arrangement test program

$L_4(2^3)$ is a two-level orthogonal table, with which the table is scheduled to run for four trials, which can arrange for up to three two-level factors. This case is three because of two-cable test, use it to arrange the test is appropriate. The arrangement of the test plan is shown in Table 1.2.4.

3. Analysis of test results

The slump of the No. 4 test   root loss results recorded in the Table 1.2.4 on the right.

Table 1.2.4 $L_4(2^3)$ Test result of extreme difference calculation

Experimental number	A. Cement varieties 1	B. Slump 2	C. Outlet temperature 3	Slump loss (cm)
1	1(A)	1(High)	1(21)	9.5
2	2(B)	1(High)	2(29)	12.7
3	1(A)	2(Low)	2(29)	4.1
4	2(B)	2(Low)	1(21)	5.1
K_1	13.6	22.2	14.6	
K_2	17.8	9.2	16.8	SUM:31.4
R	4.2	13.0	2.2	

Chapter 1 Orthogonal Design Preliminary

a) Directly to see No. 3 test slump loss less V combination of conditions, namely, a cement, the initial tree drop 8 – 9 cm, the outlet temperature of 29 ℃.

b) Calculate the sum of each factor column, calculate the corresponding level of the loss of tree height K_1 and K_2 and its range R. The calculation results are listed below Table 1.2.4. $K_1 + K_2 = 31.4$ cm Columns of each column (total slump loss of 4 tests), indicating that the calculation is correct.

From the range of the difference R, the primary and secondary order of the slump loss influenced by various factors is $B \rightarrow A \rightarrow C$, that is, the initial slump is the main factor affecting the slump loss, the cement type and the outlet temperature is secondary cause. "Calculated from the calculation of the better conditions, namely, a cement, initial slump 8 – 9 cm, 21℃ outlet temperature.

In order to minimize the slump loss of concrete, it is advisable to mix the concrete with the slump of Grade A with concrete of slump at a lower temperature.

The above two examples are the basic method of equal level orthogonal design, and the orthogonal design with different horizontal numbers will be introduced in the second chapter.

Example 1.2.1 three-factor three-level test, such as a comprehensive test required 27 times, while the orthogonal design made only 9 times; Example 1.2.2 three-factor two-level test, such as a comprehensive test required 8 Times, while orthogonal design made only 4 times. We naturally ask these 9 tests or these 4 tests can generally reflect the results of 27 trials or 8 trials? In the next section we illustrate this problem.

1.3 The Basic Principles and Characteristics of Orthogonal Design

Taking the orthogonal tables $L_4(2^3)$ and $L_9(3^4)$ as an example, we use geometry to illustrate the basic principles of orthogonal design. For those who want to know more about the problem, we can refer to References [1] [2] [3] [4] [5].

The first with $L_4(2^3)$ arranged three levels of two tests.

If you meet each of the three factors in Table 1.2.3 for a total of eight different combinations or eight different test conditions, see Table 1.3.1.

Table 1.3.1 Test combination condition

Test number	A. Cement varieties	B. Slump	C. Outlet temperature	Combination condition
1	A_1 (A)	B_1 (High)	C_1 (21)	$A_1 \, Bv_1 \, Cv_1$
2	A_1 (A)	B_1 (High)	C_2 (29)	$A_1 \, B_1 \, C_2$
3	A_1 (A)	B_2 (Low)	C_1 (21)	$A_1 \, B_2 \, C_1$
4	A_1 (A)	B_2 (Low)	C_2 (29)	$A_1 \, B_2 \, C_2$
5	A_2 (B)	B_1 (High)	C_1 (21)	$A_2 \, B_1 \, C_1$
6	A_2 (B)	B_1 (High)	C_2 (29)	$A_2 \, B_1 \, C_2$
7	A_2 (B)	B_2 (Low)	C_1 (21)	$A_2 \, B_2 \, C_1$
8	A_2 (B)	B_2 (Low)	C_2 (29)	$A_2 \, B_2 \, C_2$

Orthogonal Design in Concrete Application

The eight test numbers of Table 1.3.1 are represented by eight vertices of the cube, as shown in Figure 1.3.1.

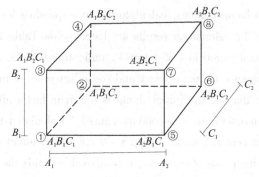

Figure 1.3.1 *The geometric figure of the 8 experiment*

According to the principle of equilibrium dispersion, we select four points ①, ④, ⑥ and ⑦ from the eight experimental points in the figure. They are evenly distributed in the square and scattered in all corners, showing as from six From a plane point of view, the four points on each side pick two points. From the twelve sides, one point is picked out from two points on the edge of the bar. By the same token, we can also choose ②, ③, ⑤, ⑧ these four points. Obviously, singled out ①, ④, ⑥, ⑦ or ②, ③, ⑤, ⑧ these four points in the square body evenly dispersed. If the ①, ⑥, ④, ⑦ The four test points are shown in Table 1.3.2, which is exactly the experimental protocol in Table 1.2.4.

Table 1.3.2 Scheme comparison of four experiment points

Test number	Factor		
	A	B	C
1①	1(A)	1(High)	1(21)
2⑥	2(B)	1(High)	2(29)
3④	1(A)	2(Low)	2(21)
4⑦	2(B)	2(Low)	1(29)

Thus, these four tests basically reflect eight experimental conditions. The test conditions of the four test numbers in Table 1.3.2 are all different and are changing. According to the table, we make the test and calculate and analyze the data obtained in this case. Some readers Q: Are there any more comparisons between them? Can we distinguish the role of each factor alone? The answer is yes. According to the principle of neat and comparability, we let A, B and C three factors in the test neatly and regularly change the changes in the comparison of the differences and the relationship between factors and levels, which is orthogonal test arrangements Cleverness. E. g:

	Slump change	Outlet temperature change
A_1 (Cement A)	B_1 (High)	C_1 (21)
	B_2 (Low)	C_2 (29)
A_2 (Cement B)	B_1 (High)	C_2 (29)
	B_2 (Low)	C_1 (21)

Chapter 1 Orthogonal Design Preliminary

It can be seen that when A takes A_1, both levels of factors B and C are changed; when A takes A_2, both levels of B and C also change. This shows that when the factor A changes from A_1 to A_2, the influences of the factors B and C are offset each other, and thus the differences between the two levels of data correspond to A_1 and A_2, which are mainly caused by the different levels. Similarly, factors B and C have similar properties. This property is called neat comparability. It is therefore able to make an intuitive analysis of orthogonal test data (range analysis), the truth is here.

In a similar way, we discuss the three-factor, three-level test with $L_9(3^4)$. Its 27 different test combinations look at Table 1.1.2. The grids of 27 test points in the table are shown in Figure 1.3.2.

According to the principle of equilibrium dispensability, we select nine test points from the 27 tests in Figure 1.3.2 and show them with "0" in the figure. These nine test points are also Table 1.2.2. A trial number of experimental programs.

The net consisting of nine points with "0" in Figure 1.3.2 gives a very vivid and intuitive representation of the balanced scatter. Because: the cube of the front, middle. After the three levels have three points left, middle and right three on the surface of the three points, the next three levels are also three points. On the case of Example 1.2.1, that is, in the slag, stone and iron on the three levels each hate the transport times. The nine test, a better representation of the 27 trials. According to the principle of neat and comparability, we analyze the factor A (slag). The effect of the three levels is to let B and C compare to A under the same changing conditions, which is

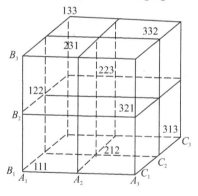

Figure 1.3.2 *The geometric figure of the 27 experiment*

	Slag content change	Iron powder change
	2%	3%
10%	3.5%	6%
	5%	9%
	2%	6%
15%	3.5%	9%
	5%	9%
	2%	9%
20%	3.5%	3%
	5%	6%

Orthogonal Design in Concrete Application

In the slag content of 10%, three kinds of stone content, three iron powder content have become; 15% and 20%, they are also three cases have changed. In this way, when calculating K_1, K_2 and K_3 of factor A, the effects of factors B and C are negated, leaving only the effect of A. Similarly, for the three types of stone mines, the amount of iron ore, the amount of iron ore, the amount of iron ore, the amount of iron ore, the amount of iron ore, and A and B are also under the same conditions, no longer described in detail here Therefore, despite the different conditions of the nine tests, the experimental arrangement was clever so that the effects of the three factors were clearly separated and their effects could be estimated to find better process conditions or formulations.

From this, it seems that if the orthogonal design, instead of using the isolated variable method as in Example 1.1.1, nine trials were selected from 27 tests, not only is the representativeness of the test point not strong, Analysis of test results, it will not have the characteristics of neat and comparable.

The orthogonal table $L_9(3^4)$ can be arranged up to four three-level factor tests. If each of the four factors is permutated and aligned, $3^4 = 81$ trials are required, and the test is conducted on an orthogonal table for only 9 attempts. Now to ask: 9 times the test can reflect the test results of 81 times? For more than three factors, and more than three for each level, it is not possible to represent the equilibrium dispersion in a geometrical way, and we can use an example of 81 test results to illustrate the equilibrium of the orthogonal design Scattered and neatly comparable.

【Example 1.3.1】 In order to investigate the impact of cement composition, cement temperature, initial temperature of the mixture and water-cement ratio on the compressive strength of concrete, reference [6] has made 81 comprehensive tests. The factors and levels of the tests and the results of their 81-intensity tests are shown in Table 1.3.3 and Table 1.3.4, respectively.

Of the 81 intensity values in Table 1.3.4, the highest intensities were 705 kg/cm and 700 kg/cm, with the combined conditions of respectively. It is $A_3 B_3 C_2 D_1$ and $A_1 B_1 C_1 D_1$ Now we set the test according to the factors and levels in Table 1.3.3 by $L_9(3^4)$ orthogonal table.

Table 1.3.3 Factor level table

Level	Factor			
	A. Cement component	B. Cement temperature F(℃)	C. Initial temperature F(℃)	D. Water cement ratio
1	1	160(71)	80(26.6)	0.40
2	9	180(82)	95(35.0)	0.47
3	7	200(93)	110(43.3)	0.54
No.	Cement component (%)					
	C_3A	C_4AF	C_3S	C_2S	$CaSO_4$	Ratio area(cm^2/g)
1	8.1	8.8	44.7	29.3	3.9	2 830
9	4.7	14.9	57.7	13.2	5.4	3 740
7	11.1	10.6	58.2	8.9	4.2	3 660

The results of the 9 experiments have been shown in Table 1.3.4. Test scheme and maximum difference calculation result column and Table 1.3.5.

From the visual analysis of the results:

Chapter 1 Orthogonal Design Preliminary

1. The major and minor order of influencing strength of each factor is $D \rightarrow A \rightarrow C \rightarrow B$ that is, the water-cement ratio is the main factor affecting the strength, and the hydrant component is the secondary factor. The influence of the initial temperature of the mixture and the temperature of the cement is small.

2. It is obtained by the value of K of each factor column that the combination condition of intensity $A_1 B_1 C_1 D_1$ (the difference of three levels is very small) is the test condition of No. 1 test.

It can be seen that the nine tests arranged in an orthogonal design can basically reflect the situation of 81 tests and are substantially the same as the conclusions of the 81 comprehensive tests. At the same time, we further see from this example that when the more factors of experimental investigation, the higher the efficiency of using orthogonal.

Table 1.3.4 81 Strength test results (kg/cm^2)

		D_1			D_2			D_3		
		B_1	B_2	B_3	B_1	B_2	B_3	B_1	B_2	B_3
C_1	A_1	700	683	685	561	568	617	483	454	490
	A_2	647	630	624	527	540	539	405	441	432
	A_3	697	689	701	612	588	595	519	501	511
C_2	A_1	656	702	621	584	598	581	444	598	475
	A_2	662	698	589	586	535	496	420	460	411
	A_3	705	681	705	607	586	627	507	496	529
C_3	A_1	693	681	611	578	615	561	479	506	482
	A_2	640	621	604	542	516	520	408	441	421
	A_3	641	633	684	524	594	620	497	508	511

Note: Aggregate: ash = 4:1, D_{max} = 19 mm, specimen size = 10 cm × 10 cm × 10 cm = 10 cm × 10 cm × 10 cm, curing at 21 degree water, each group of data is three pieces of specimen average value of 28-day dintensity.

Table 1.3.5 $L_9(3^4)$ Test result of extreme difference calculation

Experimental number	A. Cement content	B. Cement temperature	C. Mixture temperature	D. Water cement ratio	28-day Compressive strength (kg/cm^2)
	1	2	3	4	
1	1(1)	1(160)	1(80)	80(0.40)	700
2	1(1)	2(180)	2(95)	2(0.47)	598
3	1(1)	3(200)	3(110)	3(0.54)	482
4	2(9)	1(160)	2(95)	3(0.54)	420
5	2(9)	1(180)	2(110)	3(0.54)	621
6	2(9)	2(200)	3(80)	1(0.40)	539
7	2(7)	3(160)	1(110)	3(0.47)	524
8	3(7)	2(180)	1(80)	3(0.47)	501
9	3(7)	3(200)	2(95)	1(0.40)	705
K_1	1 780	1 644	1 740	2 026	
K_2	1 580	1 720	1 723	1 661	SUM : 5 090
K_3	1 730	1 726	1 627	1 403	
R	200	82	113	623	

Orthogonal Design in Concrete Application

In summary, orthogonal design has the following advantages:
1. Regularly reduce the number of tests, as a representative part of the test, and in a complicated experiment to make a scientific analysis of the results (sometimes without reducing the number of tests, The problem analysis can also be more clearly understood.)
2. The estimation of experimental error is given using the space-time difference. Orthogonal design also examines the interaction between factors and facilitates regression analysis. These two issues will be discussed in Chapters V and VI, respectively.

Chapter 2
Flexible Use Orthogonal Design

In the first chapter, we introduce the basic method of orthogonal design with equal number of levels and only one evaluation index. Concrete tests, pick the factors, the level of selection, to determine the indicators often encounter the following situations:

1. The number of a natural level of factors, it may not be exactly equal, but may not find the counterpart of the orthogonal table, or too many times to test the number.
2. The level of a factor (with disk) selection, with the level of another factor (use set) may be.
3. Cross the table of the number of columns is not enough.
4. Test pain index more than one, but several, one indicator is good, and the other is poor. To solve the above problems, make every effort to use E pay design. In this chapter, the basic methods of unequal level (mixed level), quasi level, activity level, complex factor and multi-indicator orthogonal design will be discussed with examples.

2.1 Orthogonal Design with Different Number of Levels

In concrete tests, sometimes the number of levels naturally occurs. For example, one factor is the type of cement, and the others are four types. There are four levels. The other factor is the two types of super plasticizers. They are divided into two levels. They are the level of the number of different. Even for continuously changing factors, sometimes you want to focus on a particular factor and need more levels. Others have different levels due to different factors. Encounter such problems, you can recruit a different number of horizontal orthogonal tables, referred to as the mixed level table. Appendix Ⅲ lists a variety of mixed level table, for professional workers to use. Here is an example to illustrate the number of different levels of the basic method of orthogonal design.

【Example 2.1.1】 An underground project requires the use of two types of flow concrete, No. 300 and No. 350. The so-called flow able concrete is a concrete mixture having a slump of 5 – 7 cm (Be called base concrete) Transported to the site of construction Location, then add inflow (high efficiency water reducer), mixing into the slump of 20 cm or more, easy to flow, the quality of the same concrete as the benchmark concrete. In the production of concrete, the required amount of cement does not exceed 350 kg/m^3. According to the information, mobile concrete can vibrate without compaction. In the specific conditions of the project, but also hope for less or less vibration. Whether to weaken or shorten the time to vibrate is the key problem to be studied in the experiment. At the same time, in order to improve the workability of concrete, it is proposed to add 0.25% of calcium in concrete first, and then mixed with 5% fly ash and 0.6 – 0.7 Shanghai wash oil, Through the test, determine the vibration conditions and concrete mix.

Other conditions in the test I maximum diameter of 4 cm stone, the amount of 1 100 kg/m^3, water consumption was fixed at 170 kg/m^3.

Assessment indicators: 28-day compressive strength.

2.1.1 Factors and Levels

Due to the focus on the investigation of vibration conditions, the factors selected four levels; and fly ash content, suffocated wash oil mixing, gray water ratio were selected two levels. Factors and levels are listed in Table 2.1.1.

2.1.2 Select the Orthogonal Table to Arrange the Test Protocol

The vibration conditions in this example are four levels, with the remaining three factors being two levels. This is a test of a varying number of levels (mixed levels) that requires testing with mixed level orthogonal tables. $L_8(4^1 \times 2^4)$, $L_{12}(6^1 \times 2^2)$, $L_{16}(4^4 \times 2^3)$ etc. They are all mixed level orthogonal tables. This example is the counterpart $L_8(4^1 \times 2^4)$. There are 8 crossings in the table, indicating that 8 tests should be conducted using this table. There are 5 columns, and you can arrange for a four-level factor and up to four second-level factors. This case for a full test, you need $4 \times 2 \times 2 \times 2 = 32$ times. Orthogonal design only needs to make 8 times.

Table 2.1.1 Factor level table

Level	Factor			
	A. Vibration condition (%)	B. Water cement ratio	C. Fly ash(%)	D. Wash oil content
1	Without vibration	2.0	0	0.6
2	Vibration 15 s	2.4	5	0.7
3	Vibration 30 s			
4	Pound			

Methods and procedures for arranging experiments using mixed-level orthogonal tables are the same as those for equal-level orthogonal tables and will not be repeated here. The test plan is shown in Table 2.1.2.

2.1.3 Analysis of Test Results

The strength of the eight tests is reported on the right hand side of Table 2.1.2. The calculation of the difference of the mixed level meter is the same as the equal level, and the results are listed in this table. For the convenience of visual judgment, the average value of the indexes corresponding to each factor column is given below Table 2.1.2 below (without calculation in general), and the calculation method is as follows:

For frist list

$$\overline{K}_1 = \frac{K_1}{2} = \frac{725}{2} = 363$$
$$\overline{K}_2 = \frac{K_2}{2} = \frac{945}{2} = 473$$
$$\overline{K}_3 = \frac{K_3}{2} = \frac{945}{2} = 473$$
$$\overline{K}_4 = \frac{K_4}{2} = \frac{960}{2} = 480$$

For second list

$$\overline{K}_1 = \frac{K_1}{4} = \frac{1\,623}{4} = 406$$
$$\overline{K}_2 = \frac{K_2}{4} = \frac{1921}{4} = 480$$

Table 2.1.2 $L_3(4^1 \times 2^1)$ **Test result of experiment scheme**

Experimental number	A	B	C	D		28-day Compressive strength(kg/cm^2)	Sand rate(%)
	1	2	3	4			
1	1(0 s)	1(2)	2(5)	2(0.7)	1	313	39.9
2	3(30 s)	2(2.4)	2(5)	1(0.6)	1	500	37.6
3	2(15 s)	2(2.4)	2(5)	2(0.7)	2	481	37.6
4	4(0 s)	1(2.0)	2(5)	1(0.6)	2	432	39.9
5	1(0 s)	2(2.4)	1(0)	1(0.7)	2	412	37.9
6	3(30 s)	1(2.0)	1(0)	2(0.6)	2	445	40.2
7	2(15 s)	1(2.0)	1(0)	2(0.7)	2	445	40.2
8	4(0 s)	2(2.4)	1(0)	2(0.6)	1	528	37.9
K_1	725	1 623	1 818	1 777	1 774		
K_2	914	1 921	1 726	1 767	1 770		
K_3	945					SUM:3 544	
K_4	960						
R	235	298	92	10	4		
$\overline{K}_1$	363	406	455	444	444		
$\overline{K}_2$	457	480	432	442	443		
$\overline{K}_3$	473						
$\overline{K}_4$	480						

(Institute of science and research, Ministry of communications, 00069, 1979)

The other columns are calculated in the second column.

There are two points to note: (a) For a mixed level meter, the primary and secondary order of factors should be determined according to the index and range size. determine the primary and secondary order of the factors according to the magnitude of the indicator and the magnitude of the indicator; Because the number of levels of each factor is not equal, when both factors have the same impact on the indicator, the level of the factors should be greater. (b) The gap between the mixed tables can be used in the visual analysis Estimate the experimental error for relative comparisons. To give the exact size of this error must be analyzed by analysis of variance.

Calculated from the range of Table 2.1.2:

1. The main order of influence factors is $B \rightarrow A \rightarrow C \rightarrow D$, that is, the gray – water ratio is the main factor affecting the strength. Vibration condition is an important factor: the coal ash is the secondary factor, while the water reducer agent has little effect.
2. Vibration can effectively improve the strength of the flow of concrete; therefore, the flow of concrete must be vibrated.
3. Under conditions that must be vibrated, the gray-water ratio can meet the requirement of No. 350 concrete without exceeding 2.0.
4. Adulterated with 5% fly ash, although the strength slightly decreases, it still meets the strength of design requirements.
5. Water reducer agent firefly to 0.6% is appropriate.
6. The empty column is very small; the error of this test is not big.

In summary, the fourth test meets the design requirements. Its combination condition is $A_1 B_1 C_2 D_1$ The combination of the conditions that the gray water ratio of 2.0, the vibration conditions for plugging dust, fly ash mixed; t is 5%, water reducer dose of 0.6%. The rate was 39.9%. Under the above conditions, cement is the least expensive, only 323 kg/m^3.

2.2 The Proposed Level of Law

Orthogonal table is a regular table, its number of columns, the number of levels and the number of tests have maintained a certain relationship between. After selecting the factors to be investigated, not all of the orthogonal tests at any number of levels have been combined. For continuously changing type of factors, the number of levels can be appropriate and selected according to need, the problem is not difficult to solve. For discrete factors, such as three super plasticizer, four kinds of coal ash, the number of natural formation, not Easy to change, sometimes you may not be able to find the Orthogonal Table, in this case, you need to adjust the number of levels in order to select the appropriate orthogonal table.

For example, in the mixed concrete test of flowing concrete, the factors and levels provided by the test conditions are listed in Table 2.2.1 in order to initially determine the dosage of additive agent, fly ash content and suitable sand rate. Table in the fly ash content, the amount of gravel and gray-water ratio of the three factors are three levels, while the super plasticizer species only two levels. If the level of the number of static, you need to use $L_{18}(2 \times 3^7)$ to arrange the test, but too many tests too suspect, then you can adjust the number of water reducing agent, the number of the

Chapter 2 Flexible Use Orthogonal Design

originalsuffocated wash oil and Anyang two Expand horizontally to three levels. Since it is hoped that a policy study of suffocated oil, it can be used as a third factor of the level of factors, which we call pseudo-almost, as shown in Table 2.2.1. So, you can choose $L_9(3^4)$ orthogonal table, the test less than half less.

Table 2.2.1 Factor level table

Level	Factor			
	Water reducer(0.5%)	Fly ash(%)	Gravel(kg/m³)	Water cement ratio
1	Wash oil	0	1 050	1.8
2	Anyang MF	5	1 100	2.0
3	Wash oil	10	1 150	2.2

In arranging the test protocol, in addition to the suffocate wash corresponding to the horizontal number "1", it is also necessary to correspond to the horizontal number "3" under the tap. As level 3 suffocated wash oil, known as the proposed level. Therefore, the orthogonal table of the quasi-level method is in the horizontal number of more orthogonal table; arrange the number of factors in the level of a less. Using Anyang do three tests, and suffocated oil micro-six tests. In the calculation of the very poor, respectively, to find the level of 1,3 and the third test of the sum of indicators to see how the difference with the $K_1 \setminus K_3$, if the difference is not significant, then the effect of suffocated washing oil is relatively stable and reliable; Otherwise Suffocated oil shows that with other factors with the way water played an additional role, quiet more complicated. The following combination of an example of quasi-horizontal orthogonal design analysis.

【Example 2.2.1】 Focuses on the effect of the maximum particle size of the aggregate on the compressive strength of the concrete[8]. Aggregate maximum size selection of five levels, the choice of four levels of cement dosage, sample size of two levels, which is a three-factor mixed level test. According to the above level of factors, we cannot find the orthogonal table, we can aggregate the maximum size of six levels, and factors and levels are shown in Table 2.2.2. Table A factor level "6" for the proposed level.

Table 2.2.2 Factor level table

Factor	Level					
	1	2	3	4	5	6
A. Particle side(mm)	152	76	38	19	9.5	(38)
B. Cement(kg/m³)	167	278	335	391		
C. Size(cm)	φ61×122	φ45.7×91				

Assessment indicators: 90 d compressive strength.

2.2.1 Test Program and the Results of the Very Poor Calculation

Table 2.2.2 according to the factors and levels, the table $L_{24}(6^1 \times 4^1 \times 2^3)$ can be used to arrange the test. The test plan is shown in Table 2.2.3.

Orthogonal Design in Concrete Application

According to the factor level of Table 2.2.2, table $L_{24}(6^1 \times 4^1 \times 2^3)$ can be used to arange the experiment, and the test plan is listed in Table 2.2.3. The factor A in the table corresponds again to the horizontal number "6" in addition to the maximun aggregate size of 38 min corresponding to the hori zontal number "3". Level 1,2,4,5 owere tested four times each, and level 3 was tested eight times. The test results were recorded on the right side of Table 2.2.3 results were recorded on the right side of Table 2.2.3. The results of its vange calculation are shown in table 2.2.4, $\overline{K}_1, \overline{K}_2, \overline{K}_4, \overline{K}_5$ in the table are the average value of the four test results. and $\overline{K}_3$ is the average of the eight test results. Because $\overline{K}_3$ is almost egual to $\overline{K}_6$.

2.2.2 Intuitive Analysis of Test Results

From the range calculation results shown in Table 2.2.4:

1. The main factors that influence the use of cement R_{90} are the following factors: the maximum aggregate size is a secondary factor; and the influence of specimen size is small, within the experimental error range.

Table 2.2.3 $L_{24}(6^1 \times 4^1 \times 2^3)$ Test result of experiment scheme

Experimental number	A	B	C			90-day Compressive strength(kg/cm^2)
	1	2	3	4	5	
1	1(152)	1(167)	1(61×122)	1	2	229
2	1	2(278)	1	2	1	297
3	1	3(335)	2(45.7×91)	2	2	290
4	1	4(391)	2	1	1	335
5	2(76)	1	2	1	1	174
6	2	2	2	1	2	293
7	2	3	1	1	1	340
8	2	4	1	2	2	349
9	3(38)	1	1	1	1	137
10	3	2	1	2	2	349
11	3	3	2	2	1	333
12	3	4	2	1	2	372
13	4(19)	1	2	2	2	120
14	4	2	2	1	1	295
15	4	3	1	1	2	309
16	4	4	1	2	1	339
17	5(9.5)	1	1	1	1	93
18	5	2	1	2	2	228
19	5	3	2	2	1	293
20	5	4	2	1	2	355
21	6(38)	1	2	2	2	139
22	6	2	2	1	1	327
23	6	3	1	1	2	326
24	6	4	1	2	1	337

Chapter 2 Flexible Use Orthogonal Design

Table 2.2.4 Test result of 90-day compressive strength

	A	B	C			The sum of the strength of the 24 test
	1	2	3	4	5	
$\bar{K}_1$	1151(288)	892(150)	3 267(272)	3 411(284)	3 300(275)	6 593
$\bar{K}_2$	1 156(289)	1 723(287)	3 326(277)	3 182(265)	3 293(274)	
$\bar{K}_3$	1 125(282)	1 891(315)				
$\bar{K}_4$	1 063(266)	2 087(348)				
$\bar{K}_5$	969(242)					
$\bar{K}_6$	1 129					
$\bar{R}$	187(47)	1 195(198)	59	229	7(1)	

NOTE: The difference in the mean value or the mean value in the parenthesis.

2. In the case of a certain amount of cement and specimen size, the strength R_{90}, With the largest aggregate size decreased. The largest particle size in the range of 152 – 38 mm, R_{90} Basically similar; maximum size of 19 mm and 9.5 mm of small aggregate concrete, Lower by 5% and respectively.

3. The strength R_{90}, increases with the cement, although K * is almost doubled when the cement S is increased from 167 kg/m^3 to 278 kg/m^3. If the dosage is increased, the strength will be increased.

2.3 Activity Level and its Application

In Orthogonal Design, Level selection of a factor the water content of a factor L (using 1) depends on the level of another factor; or it is known that there is a certain degree of Dependencies. At this time, the level of activity can be used to arrange the test program.

Look at a simple example. Select concrete admixture agents and agents, taking two factors, the two levels are shown in Table 2.3.1 a.

Table 2.3.1a Factor level table

Level	Factor	
	Type	Content (percentage of cement consumption)
1	Aerated agent	Less
2	Water reducer	More

According to the different dosage of agents to arrange the number of agents, this approach is called activity level (see Table 2.3.1 b). If you use a fixed level (dose), the dose of air entraining agent is not suitable for reducing the amount of wood calcium, with Cheung; wood clever agent firefly is not suitable for air entraining agent. This kind of experiment can be arranged at the level of activity.

Orthogonal Design in Concrete Application

Table 2.3.1b   Activity level of admixture

Content	Aerated agent	Water reducer
Level 1 (low)	$0.8/10^4$	0.20%
Level 2 (high)	$1.0/10^4$	0.25%

Take the test of frost resistance of fly ash mortar as an example to illustrate this kind of experiment.

【Example 2.3.1】 According to some domestic and foreign tests on the frost resistance of concrete, it is considered that the frost resistance of concrete decreases significantly with the increase of the ashes. For the touch of the sample with fly ash concrete frost resistance performance, the Ministry of Transportation Air Force Bureau used two times $L_9(3^4)$ orthogonal table, made 18 tests, respectively, examine the content of coal ash, the fineness of ordinary mortar and gas Mortar frost resistance is now based on the original test data $L_{12}(3^1 \times 2^4)$, the activity level is arranged in an orthogonal table, only using 12 test results, you can more clearly explain the problem.

Assessment index: freezing thawing loss rate.

2.3.1   The Factors and Levels are Shown in Table 2.3.2

Table 2.3.2   Factor level table

Level	Factor		
	A Fly ash(%)	B Type of mortar	C Fineness
1	15	Ordinary mortar	Fine
2	30	Aerated mortar	Coarse
3	40		

Table 2.3.2 fineness due to turbulence is divided into two levels of thinner and coarser, the fine level is actually 4% and 8.6% of two kinds of fineness, the coarser level actually has 6.6% and 15.4% two kinds of fineness. We arrange the fineness-degree water according to the different types of mortar: the specific fineness of the flatness, as shown in Table 2.3.3. The fineness level varies with the type of mortar in the table as "activity level".

Table 2.3.3   Fineness level table

Fineness	Type	
	Level 1 Ordinary mortar	Level 2 Aerated mortar
1 (fine)	4%	6.6%
2 (crude)	8.6%	15.4%

According to the factors and levels, choose $L_{12}(3^1 \times 2^4)$ orthogonal 2.3.2 Test Table Arrangement is Listed in Table 2.3.4.

Chapter 2 Flexible Use Orthogonal Design

Table 2.3.4 $L_3(3 \times 2^4)$ Test result of experiment scheme

Experimental number	A. Content 1	B. Type 2	C. Fineness 3	4	5	Weight loss rate of 25 freeze-thaw cycles(%)
1	2(30%)	1(Pu)	1(4%)	1	2	2.00
2	2(30%)	2(Jia)	1(6.6%)	2	1	0.112
3	2(30%)	1(Pu)	1(8.6%)	2	2	8.500
4	2(30%)	2(Jia)	2(15.4%)	1	1	0.045
5	1(15%)	1(Pu)	1(4%)	2	2	1.300
6	1(15%)	2(Jia)	1(6.6%)	2	1	0.042
7	1(15%)	1(Pu)	2(8.6%)	1	1	2.100
8	1(15%)	2(Jia)	2(15.4%)	1	2	0
9	3(40%)	1(Pu)	1(4%)	1	1	8.4
10	3(40%)	2(Jia)	1(6.6%)	1	2	0.189
11	3(40%)	1(Pu)	2(8.6%)	2	1	45.300
12	3(40%)	2(Jia)	2(15.4%)	2	2	0.573
K_1	3.442	67.6	12.043	12.734	55.999	SUM:68.561
K_2	10.657	0.961	56.518	55.827	12.562	
K_3	54.462					
R	51.020	66.639	44.475	43.093	43.437	

In the tests No. 1 and No. 2 in Table 2.3.4, although the factor of fineness is both Level 1 (finer) and the value of the fineness is different from the value of Low, in Test No. 1, The mortar type is plain sand violet fineness is 4%; in the No. 2 test, the fineness is 6.6% because the mortar type is gas mortar. The same is true for the six tests of levels 2 for fineness factors in No. 5, 6 and 9, 10.

2.3.2 Experimental Results of the Analysis

The result of the 12 tests and the calculation of the range difference are recorded respectively in the right and left sides of Table 2.3.4. From the very poor size can be drawn:

1. The major and minor bottlenecks that affect the frost resistance are $B \rightarrow A \rightarrow C$, that is, the type of mortar is the main factor influencing frost resistance. Fly ash doping is the second factor, but the effect of fineness is small.
2. Aerated mortar can significantly improve its frost resistance.
3. Frost resistance of mortar decreases with increasing fly ash addition. When the firefly from 30% to 40%, its frost resistance decreased significantly.
4. Considering the fineness is not the main factor, the fineness is too fine, the cost is increased, and the process is difficult, so no more than 8.6% is appropriate. For example, No. 4 test, aerated mortar, fineness of 15.4%, weight loss rate of only 0.045%.

In summary, in order to improve the freeze-thaw resistance of mortar (concrete) mixed with fly ash, it is advisable to use fly ash with a fineness of not more than 8.6% and to prepare aerated mortar (concrete) at a content of 30% of.

2.4 Composite Factors and Their Applications

In orthogonal design, there are times when there are more factors to consider, but because of the limited number of tests or the number of columns that I do not have enough. At this time, several factors can be combined into one because of the inspection.

【Example 2.4.1】 In order to obtain strong cement, the Chinese Academy of Sciences, Institute of Chemical and Chemical Silicate clinker cement composition on the net strength of the impact of the broader market test. The original test data are arranged orthogonal test design and analysis program to optimize the high-strength cement mineral composition.

Assessment indicators: specimen size 2 cm × 2 cm × 2 cm of net strength.

2.4.1 Factors and Levels Listed in the Table

The table above is a composite table of factors. Table C_3S_2: C_2S constitute a factor, different proportions constitute the different levels of the factor; $C_3S_2 + C_2S$also constitute a due purple, although the total difference to become a different level of the factors and so on. By combining the total amount and the ratio of each test number, the amount of each component can be found.

Table 2.4.1 Factor level table

Factor	Level			
	1	2	3	4
A. $C_3SC_sS(\%)$	5:1	8:1	18:1	—
B. $C_2S + C_2S(\%)$	70	80	85	90
C. C_3A: $C_2AF(\%)$	0.8:1	0.5:1	—	—

2.4.2 Select the Orthogonal Table to Arrange the Test Protocol

In this case, the orthogonal $L_{24}(3^1 \times 4^1 \times 2^1)$ is that it is also a mixed level orthogonal table, with which 24 tests can be performed, which can arrange for a three-level factor, a four-level factor and a maximum of four Horizontal factors. The test scheme and test results are shown in Table 2.4.2.

2.4.3 Analysis of Test Results

1. Looking directly at No. 14 test strength of up to 2 270 kg/cm^2, the group conditions will be $A_2B_2C_2$.
2. Calculate the sum of the horizontal and vertical scales of each factor for this mixed level and the difference between them. Example 2.3.1. The calculation results are shown in the lower part of Table 2.4.2. From the very poor size:

Chapter 2 Flexible Use Orthogonal Design

a) $C_3S:C_2S$ and $C_3A_2:C_4AF$ are the main factors affecting the strength of the paste; $C_3S + C_2S$ is a secondary factor.
b) Net strength increases with increasing $C_3S:C_2S$ ratio. When $C_3S:C_2S$ value increased from 5 to 8, the strength increased greatly; and from 8 to 18, the increase in tension was small.
c) The net strength increases with the increase of the total capacity of $C_3S + C_2S$ When the $C_3S + C_2S$ value increased from 75% to 80%, the strength increased greatly, while from 80% to 90%, the strength increased little.

Table 2.4.2 $L_{24}(3^1 \times 4^1 \times 2^1)$ Test result of experiment scheme

Experimental number	Factor			4	5	6	28-day Compressive strength (kg/cm^2)
	$C_3S:C_2S$	C_3S+C_2S	$C_3A:C_4AF$				
	Column number						
	1	2	3				
1	1(5:1)	1(75)	1(0:8:1)	1	1	1	1 840
2	1	2(80)	1	1	2	2	1 860
3	1	3(85)	1	2	2	1	1 830
4	1	4(90)	1	2	1	2	1 890
5	1	1	2(0:5:1)	2	2	2	1 850
6	1	2	2	2	1	1	2 150
7	1	2	2	1	1	2	1 896
8	1	4	2	1	2	1	2 160
9	2(8:1)	1	1	1	1	2	2 000
10	2	2	1	1	2	1	1 860
11	2	3	1	2	2	2	2 040
12	2	4	1	2	1	1	2 185
13	2	1	2	2	2	1	2 040
14	2	2	2	2	1	2	2 270
15	2	3	2	1	1	1	2 170
16	2	4	2	1	2	2	2 193
17	3(18:1)	1	1	1	1	2	2 010
18	3	2	1	1	2	1	2 090
19	3	3	1	2	2	2	2 152
20	3	4	2	2	1	1	2 133
21	3	1	2	2	2	1	2 172
22	3	2	2	2	1	2	2 230
23	3	3	2	1	1	1	2 177
24	3	4	2	1	2	2	2 090
K_1	15 476	11 912	23 890	24 346	24 951	24 807	
K_2	16 758	12 460	25 398	24 942	24 337	24 481	
K_3	17 054	12 265					SUM:49 288
K_4		12 651					
R	1 578	739	1 508	596	614	326	

d) According to the above situation, the optimal combination condition is the combination condition of No. 14 test, $A_2B_2C_2$ namely $CsS:C_2S = 8$, $C_3S + C,S = 80\%$ and $C_3A:C_4AF = 0.5$. According to the relationship between the above values and $C_2S + C_2S + C_3A + C_4AF = 100\%$, $C_3S = 70\%$, $C_2S = 10\%$, $C_3A = 7\%$ and $C_4AF = 14\%$.

2.5 Multi-indicator Orthogonal Design Analysis

In the examples cited above, there is only one assessment index, such as strength or slump or freeze-thaw loss rate, which is called the orthogonal design of single index. In some concrete tests, there is often more than one indicator used to test the effectiveness of the test, but only a few. For example, to simultaneously assess the strength and slump, etc., this type of problem is called orthogonal design of multiple indicators. In a multi-indicator test, sometimes an indicator (such as strength) is good; another indicator (such as slump) may not meet the design requirements, which is a problem that we may encounter in our practical work. How to take into account the indicators, find out the indicators are as good as possible combination of conditions? In other words, how to analyze the test results of multi-indicator orthogonal design? The following examples to introduce the "comprehensive balance of law" and "efficacy coefficient method."

2.5.1 The Comprehensive Balance of Law

Comprehensive balance method is to separate the indicators by a single indicator for analysis, and then the calculation and analysis of the results of each indicator, a comprehensive balance of the final conclusion.

【Example 2.5.1】 An underground project requires the preparation of 600 concrete, slump requirements of more than 10 cm, and the use of cement is Datong 500 ordinary cement (real measured 28-days hard intensity of 613 kg/cm^2), In order to select the mix of cmcrote with large collapsing concrete mix ratio with strength not less than the cement hardness mark, water absorber should be added to the concrete. There are three types of water reducers: (1) Anyang MF; (2) BeiJing AnYang factory MF; (3) Shanghai sulfated wash oil. Test optimization of concrete mix. Experimental conditions: gravel bone. Material, the largest particle size 20 mm > fineness modulus of river sand $FM = 3.3$ concrete Design capacity of 2 350 kg/m^3.

Assessment indicators: 28 days compressive strength and slump, the test in two batches. The first batch of experiments is to purify: (1) gray-water ratio, water consumption and the range of application of gravel; (2) the effect of the agent on the indicator at the same dosage of various super plasticizers.

Ⅰ. Test program and test results

Factors and levels are shown in Table 2.5.1.

Chapter 2 Flexible Use Orthogonal Design

Table 2.5.1 Factor level table

Level	Factor			
	A. Water cement ratio	B. Water content (kg/m³)	C. Gravel(kg/m³)	D. Water reducer (The dose was 1%.)
1	3.0	140	1 100	AN MF
2	3.5	155	1 150	JING MF
3	4.0	170	1 200	HU oil

The test plan and test results are shown in Table 2.5.2.

II. Analysis of test results

This test is the strength and slump two indicators, as mentioned earlier, first, respectively, according to a single indicator of strength and slump analysis, and then they still analyze the results of a comprehensive balance.

Table 2.5.2 $L_{24}(3^4)$ Test result of experiment scheme

Experimental number	Factor				Assessment Index	
	A. Water cement ratio	B. Water content (kg/m³)	C. Water cement ratio	D. Water reducer	Slump (cm)	28 day Compressive strength(kg/cm²)
	Column number					
	1	2	3	4		
1	1(3.0)	1(140)	3(1 200)	2(JingMF)	0.7	577
2	2(3.5)	1	1(1 100)	1(An MF)	0.4	617
3	3(4.0)	1	2(1 150)	3(Huoil)	0	636
4	1	2(155)	2	1	3.3	630
5	2	2	3	3	0.8	606
6	3	2	1	2	0	634
7	1	3(170)	1	3	7	589
8	2	3	2	2	12.3	610
9	3	3	3	1	0.6	558
			SUM:25.1			5 557

The range of test results is shown in Table 2.5.3.

Table 2.5.3 Test result of experiment scheme

	Slump(cm)				28 day Compressive strength(kg/cm²)			
	A	B	C	D	A	B	C	D
K_1	11.0	1.1	7.4	4.3	1 796	1 830	1 840	1 905
K_2	13.5	4.1	15.6	13.0	1 833	1 870	1 876	1 821
K_3	0.6	19.9	2.1	7.8	1 928	1 857	1 841	1 831
$\overline{K_1}$	3.7	0.4	2.5	1.4	599	610	613	635
$\overline{K_2}$	4.5	1.4	5.2	4.3	611	623	532	607
$\overline{K_3}$	0.2	6.6	0.7	2.6	643	619	614	610
$\overline{R}$	4.3	6.2	4.5	2.9	44	13	12	28

(1) As can he seen from Table 2.5.3, the main factor affecting the collapse dgree is water quantity, followed by the amount of stones and the water cement ratio, little effect on the species. The combination of conditions to achieve higher slump is $A_2B_3C_2D_2$. (2) The main factor influencing the intensity is the ratio of gray water to water, followed by the seed, and the water consumption and gravel use have little effect. The combination of conditions that satisfy the strength requirement is $A_3B_2C_2D_1$ (or D_3).

Comprehensive slump and strength of the combination of both conditions, the initial selection of gray water than 3.5 - 4.0, water 155 - 170 kg/m³, stone dosage 1 150 kg/m³, Anyang MF or Shanghai suffocation oil.

The purpose of the second batch of experiments is: on the basis of the first batch of experiments, narrow the scope of the test, and appropriately increase the dosage of naphthalene super plasticizer, and at the same time, mix and mix with calcium, calcium and water to further clarify the increase of the motility of the warts The effect of strength. To economic and reasonable choice of mix.

Ⅰ. The factors and levels are shown in Table 2.5.4

Table 2.5.4　Factor level table

Level	Factor					
	A. Water cement ratio	B. Gravel	C. Water content	D. Type	E. Content	F. Ca
1	3.5	1 140	155	HU oil	1%	0
3	4.0	1 160	170	AN MF	1.5%	0.2%

Ⅱ. Use orthogonal table arrangement test program

Orthogonal table $L_8(2^7)$ There are 7 columns in the orthogonal table. Eight tests are required to use this table. It can arrange up to seven second-level factors. This example has six second-level factors that can be used to schedule the test. This example for a full test, to be $2^6 = 64$ times, orthogonal design for only 8 times, indicating that the more factors to consider, the orthogonal design efficiency (reduce the number of trials) the more times. The test protocol is shown in Table 2.5.5. The third column in the table is empty, as an estimate of the experimental error

Ⅲ. Analysis of test results

Table 2.5.5　$L_8(2^7)$ Test result of experiment scheme

Experimental number	Factor							Slump (cm)	28-day Compressive strength (kg/cm²)
	A	B		C	D	E	F		
	Column number								
	1	2	3	4	5	6	7		
1	1(3.5)	1(1 140)	1	2(170)	2(AN AF)	1(1.0)	2(0.2)	6.9	560
2	2(4.0)	1(1 160)	1	2	1(Huoil)	1	1(0)	0	623
3	1	2	2	2	2	2(1.5)	1	17.9	540
4	2	2	1	2	1	2	2	2.7	581
5	1	1	2	1(155)	1	2	2	11.1	632
6	2	1	1	1	2	2	1	1.1	645
7	1	2	1	1	1	1	1	2.3	622
8	2	2	2	1	2	1	2	0	500
SUM:								42.0	4 703

Chapter 2 Flexible Use Orthogonal Design

The slump and strength of the eight trials are shown on the right in Table 2.5.5. The difference calculation results are shown in Table 2.5.6.

1. It can be directly seen from Table 2.5.5 that No. 5 test can meet the requirements of setting fall of 11.1 cm, compressive strength of 632 kg/cm². The combined conditions are: ash to water ratio: 3.5, stone dosage: 1 140 kg/m³, water 155 kg/m³ naphthalene series: Shanghai sulfonated oil. Mu-Ca:0.2%.

Table 2.5.6 Test result of range

	A	B		C	D	E	F
			Slump(cm)				
K_1	38.2	19.1	(13.0)	14.5	16.1	9.2	21.3
K_2	3.8	22.9	(29.0)	27.5	25.9	32.8	20.7
$\bar{K}_1$	9.6	4.8	(3.3)	3.6	4.0	2.3	5.3
$\bar{K}_2$	1.0	5.7	(7.3)	6.9	6.5	8.2	5.2
R	8.6	0.9	(4.0)	3.2	2.5	5.9	0.1
K_1	2 354	2 460	(2 406)	2 399	2 458	2 305	2 430
K_2	2 349	2 243	(2 295)	2 304	2 245	2 398	2 278
$\bar{K}_1$	589	615	(602)	600	615	576	608
$\bar{K}_2$	587	561	(574)	576	561	600	568
R	2	54	(28)	24	54	24	40

2. From the range analysis of Table 2.5.6, it can be concluded that: (a) When the ratio of gray water to that of naphthalene super plasticizer is the main factor affecting the slump when the ratio of gray water to water is 3.5 – 4.0, The effect of using stones, water firefly, naphthalene-based agents and wood-calcium mixing is very small, all within the experimental error. More slump conditions: take A is A_1, E is E_2. (b) The main influencing factors of strength are Naphthalene series and pebbles. The secondary factor is the content of wood calcium, the ratio of gray water to water, the dosage of water and the dosage of naphthalene are all very small, all within the experimental error. Higher intensity conditions: take B for B_1, D for A, F for F_1.

To sum up, the good conditions for the selection of slump, $A_1B_1C_1D_1E_2F_1$ and strength at the same time are as follows: the ratio of gray to water is 3.5, the water consumption is 155 kg/m³ (the amount of cement is 543 kg/m³), and the amount of gravel is 1 140 kg/m³ < sand rate of 33%), Shanghai suffocation wash oil dose of 1.5%. This is the 5th test number except for the experimental conditions of calcium.

2.5.2 The Efficiency Coefficient Method

Assuming the orthogonal design of the original assessment of the nine indicators, each indicator has a certain coefficient of effectiveness, the i index of the coefficient of effectiveness $d_i (0 < d_i < 1)$, if there is n indicators, there are n eigenvalues $(i = 1, 2, \cdots, n)$, using the geometric quadrature of these coefficients to get a total efficiency coefficient

$$d = \sqrt[n]{d_1 d_2 \cdots d_n}$$

Here, the coefficient d is used to denote the total translation performance of n subscripts. In this way, after each test, as long as the coefficient d, that is the result.

The method for determining the coefficient d_i is as follows, using $d_i = 1$ to denote the effect of the index, and $d_i = 0$ means that the it index has the worst performance d_i value satisfaction.

$$0 \leqslant d_i \leqslant 1$$

Obviusly, if a test result makes tle efficacy coefficient $d_i (i = 1, 2, \cdots, n) 1$, and then the total efficiency coefficient is $d = \sqrt[n]{1 \times 1 \cdots 1} = 1$. This shows that the overall effect is good. Conversely, if there is a certain heart $d_i = 0$, then $d = 0$, that is, this The test results are not good. Therefore, in the analysis of multiple indicators, the military test is not a single indicator, only the total efficiency coefficient of a unified n indicators, so that the result analysis is greatly simplified.

【Example 2.5.2】 An engineering option to sugarcane honey tank (A) Body, composite aerated surfactant (B)——ethylene oxide fatty alcohol roasted and Sui condensate type (C)——a sodium hexametalate composition; deer concrete new water retarder "3FG", choose the best blend of each component:

Halogen levels and test results are shown in Table 2.5.7 and Table 2.5.8, respectively.

Table 2.5.7 Factor level table

Factor	Level		
	1	2	3
A. Alcohol tank	0.15	0.20	0.25
B. Condensate	0.016	0.012	0.008
C. Six polyphosphate	0.02	0.03	0.01

Table 2.5.8 $L_9(3^4)$ Calculation results of test results and range of efficiency coefficients

Experimental number	A	B	C	4	Experimental result				Total Efficiency Index $d = \sqrt{d_1 d_2}$
	1	2	3		Water reduction rate (%)	28-day Compressive strength (kg/cm²)	Water reduction rate d_1	Strength d_2	
1	1(0.15)	1(0.0016)	1(0.02)	1	15.3	312	0.83	0.91	0.87
2	1	2(0.012)	2(0.03)	2	15.3	340	0.83	0.99	0.91
3	1	3(0.008)	3(0.01)	3	14.3	314	0.78	0.91	0.84
4	2(0.20)	1	2	3	17.9	290	0.97	0.84	0.90
5	2	2	3	1	18.4	330	1.00	0.96	0.98
6	2	3	1	2	15.8	344	0.86	1.00	0.93
7	3(0.25)	1	3	2	18.4	287	1.00	0.83	0.91
8	3	2	1	3	17.9	328	0.97	0.95	0.96
9	3	3	2	1	17.3	319	0.94	0.93	0.93
K_1	2.62	2.68	2.76	2.78					SUM: 8.23
K_2	2.81	2.85	2.74	2.75					
K_3	2.80	2.70	2.73	2.70					
R	0.19	0.17	0.03	0.08					

In this case, the assessment index is the water reduction rate and the compressive strength. According to the "comprehensive balance method", the result is $A_2B_2C_0$ (C_0 is any level of the factor).

The above two indicators into a single indicator, that is, the efficacy coefficient analysis of the trial results, the total efficiency coefficient and its range calculation results are shown in Table 2.5.8.

From Table 2.5.8, it can be seen directly that the coefficient of efficacy d of No. 5 test is 0.98. The combination condition is $A_2B_2C_3$. The optimal combination condition obtained by the visual analysis results is as follows $A_2B_2C_1$. Since the influence of factor C on the efficiency factor is within the experimental error, the preferred result is $A_2B_2C_0$, which is in good agreement with the conclusion of the "integrated balancing approach." Thus, when the assessment of more than two indicators, the coefficient of efficacy analysis of test results is simpler.

It should be pointed out here that most of the two or more indicators in the concrete test are consistent, that is, the indicators are as good as possible, but sometimes they also encounter contradictory conditions, that is, an indicator is as good as possible and The other indicator is as low as possible. At this time can not use the efficiency coefficient method for optimization, the ratio of two indicators can be used to analyze the test results.

Chapter 3
Variance Analysis of Orthogonal Design

In the orthogonal analysis of the orthogonal design, we use the method of analysis of the test data of the very poor method, that is, the size of each factor column shows the size of the various factors that affect the order of the primary and secondary indicators; the best level of the factors (The largest K value) as the optimal process conditions or formula; with empty column tolerance that experimental error. The very poor method of calculating the fish is small, simple and easy, you can visually describe the problem, so it is easy to promote an analytical method. However, the very difference method does not strictly distinguish between data fluctuations caused by changes in test conditions during the test and data fluctuations caused by experimental errors. Nor does it provide a criterion for judging the effects of the factors under investigation Is significant? In order to make up for these deficiencies (poor view) analysis, analysis of variance can be used. This chapter will combine professional examples to introduce the basic concepts of variance analysis, methods and analysis of variance of orthogonal design, and discuss the error handling.

3.1 Introduction of Variance Analysis

3.1.1 The Quantitative Representation of Variation
First you must understand the quantitative representation of variation.

There are n uneven data $x_1, x_2, \cdots, x_n$, between them the difference is called variation. There are two ways to quantify the variation:

1. Range method. Which is

$$R = \max(x_1, x_2, \cdots, x_n) - \min(x_1, x_2, \cdots, x_n) \tag{3.1.1}$$

max and min mean maximum and minimum, respectively. R has been used in the intuitive analysis of orthogonal designs.

2. Sum of squares method. The disadvantage is that the poor data provided by the use of confidence is not enough. Another way to express variation is the sum of squares of variances, referred to as the sum of squares for short, and recorded as

$$S = \sum_{i=1}^{n} (x_i - \bar{x})^2 \tag{3.1.2}$$

$$X = \frac{1}{n} \sum_{i=1}^{n} x_i \tag{3.1.3}$$

The sign $\sum$ is the meaning of summation, indicating that X_1 is added straight to X_n. S is a measure of how far away the data is from the average, the larger the S, the greater the difference between the data.

【Example 3.1.1】 Six compressive strengths of concrete with the same K ratio were obtained, and the values were 217, 224, 227, 231, 236 and 258 kg/cm², respectively value.

From equations 3.1.2 and 3.1.3, find:

$$\overline{X} = \frac{1}{6}(217 + 224 + 227 + 231 + 236 + 258) = 232.2$$

$$S = (217 - 232.2)^2 + (224 - 232.2)^2 + \cdots + (258 - 232.2)^2 = 1\,006.83$$

Can be seen that the calculation of S is more troublesome. This is due to the fact that the number of $\overline{X}$ significant digits has been increased, and the computational effort has greatly increased: at the same time as the result of the When calolating S, every Squre sum has increased the error addition is rounded off, the error is added to the sum of the terms. When the data is more, this error can not be accumulated together. Therefore, the formula 3-1-2 expanded:

$$S = \sum_{i=1}^{n} (X_i - \overline{X})^2$$

$$= \sum_{i=1}^{n} x_i^2 - 2\overline{X} \sum_{i=1}^{n} x_i + \sum_{i=1}^{n} \overline{X}^2$$

$$= \sum_{i=1}^{n} X_i^2 - z \cdot \frac{1}{n} (\sum_{i=1}^{n} x_i^2)^2 + n \cdot \frac{1}{n^2} \times (\sum_{i=1}^{n} x_i^2)^2 \qquad (3.1.4)$$

$$= \sum_{i=1}^{n} x_i^2 - \frac{1}{n} (\sum_{i=1}^{n} x_i^2)2$$

Through the formula 3-1-4

$$S = 217^2 + 224^2 + \cdots + 258^2 - \frac{1}{6} \times (217 + 224 + \cdots + 258)^2$$

$$= 324\,415 - \frac{1}{6} \times 1\,393^2 = 1\,006.83$$

Although this calculation error is small, but the workload is still large. To this end, often used under Column approach:

1. Each data minus (plus) go to the same number a, square and S still constant.

In this example, if every data is subtracted by 200, ie, $x'_i = x_i - 200$, the difference x'_i is calculated as follows:

$$\sum x'_i = 17 + 24 + 27 + 31 + 36 + 58 = 193$$

$$\sum x'^2_i = 17^2 + 24^2 + 27^2 + 31^2 + 36^2 + 58^2 = 7\,215$$

$$\overline{X} = 200 + \overline{X}' = 200 + \frac{193}{6} = 232.2$$

$$S = S' = \sum_{i=1}^{n} x'^2_i - \frac{1}{n}(\sum x'_i)^2 = 1\,006.83$$

This is exactly the above result.

Chapter 3 Variance Analysis of Orthogonal Design

2. Each data with the same (by) a number b, the corresponding square and S reduced (increase) b^2 times.

In this example $x_i'' = \dfrac{x_i - 200}{10}$, The corresponding data is changed to 1.7, 2.4, 3.1, 3.6, 5.8.:

$$\sum x_i'' = 1.7 + 2.4 + 2.7 + 3.1 + 3.6 + 5.8$$
$$= 19.3$$
$$\sum x_i''^2 = 1.7^2 + 2.4^2 + 2.7^2 + 3.1^2 + 3.6^2 + 5.8^2 = 72.15$$
$$\overline{X} = 200 + 10 \times \overline{X}'' = 200 + 10 \times \dfrac{19.3}{6} = 232.2$$
$$S = 100 S'' = 100 \left[\sum x_i''^2 - \dfrac{1}{6} \left(\sum x_i'' \right)^2 \right] = 1\,006.83$$

Can be seen, the use of these two methods, so that the calculation of firefly greatly reduced. In the analysis of variance to often use, be sure to master.

【Example 3.1.2】 From the same mix of concrete, and then obtain the letter of loss of value: 214, 223, 241, 249, together with the three data in Example 3.1.1, a total of ten data, Find the average and square sum.

The original data with minus 200 divided by 10, $x_i'' = \dfrac{x_i - 200}{10}$ that is

$$\overline{X} = 200 + 10 \times \dfrac{1}{10} \sum x_i''$$
$$= 200 + 10 \times (1.7 + 2.4 + 2.7 + 3.1 + 3.6 + 5.8)/10 = 232.0$$
$$S = 100 S''$$
$$= 100 \times [1.7^2 + 2.4^2 + 2.7^2 + 3.1^2 + 3.6^2 + 5.8^2$$
$$- \dfrac{1}{10} \times (1.7 + 2.4 + 2.7 + 3.1 + 3.6 + 5.8)^2] = 1\,782$$

The mean value of 232.0 for Example 3.1.2 is very similar to the average of 232.2 for Example 3.1.2, but the sum of squares is significantly larger. From the same process conditions and the same mix to obtain six concrete strength data and ten strength data fluctuations in amplitude is about the same, why the difference between the sum of squares? From the calculation we can see that, although the six data and the amplitude of the volatility of ten data, but the calculation of the sum of squares, the former is the sum of six squares, the latter is the sum of ten squares, so the latter than the former Big. This is the same degree of volatility; the data is more than the sum of square and less than the sum of squares. Therefore, only the sum of squares to reflect the size of the volatility is not enough. In order to eliminate the influence of the number of data on the square sum, we need to introduce the concept of freedom.

The degree of freedom is denoted by f. Its definition is the number of independent variables, that is, the number of variable values minus the number of constraints received. If If a Vaviable is used to compute a parameter or statisticare n values, (then the variable only $w - 1$ values remain independent), that is, after finding a value, the last value must be calculated from the parameters or The value of the statistics value. For example, A sample with a capacity of n, x_1, x_2, $\cdots$, x_n. what is the square of the degree of freedom $S = \sum_{i=1}^{n}(X_i - \overline{X})^2$ and the degree of freedom of a sample whose volume is n number-centers? It calculates the sample mean from n values $\overline{X}$ of $\overline{X} = \dfrac{1}{n}\sum_{i=1}^{n}(x_i - \overline{X})$, that is a constraint. With this constraint, the degree of freedom is $n - 1$. Therefore, the degree of freedom of the sum of squared deviations is $n - 1$. In the analysis of square variance, there is a simple method to determine the degree of freedom, the reader here as long as the superficial understanding of the concept of freedom on the line.

Now introduce the concept of the average square sum (referred to as the mean square), the so-called mean square is almost square and divided by the corresponding degree of freedom f, the mean square with $\overline{S}$, then

$$\overline{S} = \frac{S}{f} \tag{3.1.5}$$

$$\overline{\sigma} = \sqrt{\frac{S}{n-1}} \tag{3.1.6}$$

For example 3.1.1

$$\overline{S} = \frac{S}{5} = 0.2 \times 1\,006.83 = 201.366$$

$$\hat{\sigma} = \sqrt{\overline{S}} = \sqrt{201.366} = 14.19 \text{ kg/cm}^2$$

For example 3.1.2

$$\overline{S} = \frac{S}{9} = \frac{1\,782}{9} = 198$$

$$\hat{\sigma} = \sqrt{\overline{S}} = \sqrt{198} = 14.07 \text{ kg/cm}^2$$

It can be seen that the $\overline{S}(\hat{\sigma})$ of the six intensity data and the ten intensity data are similar, which is in agreement with the same process conditions and the same compounding ratio, which shows that it is more reasonable to use to reflect the fluctuation.

To sum up, variance (or mean variance) can reflect the degree of dispersion of random variation. When C is constant, the variance has the following properties:

a) $D(C) = 0$
b) $D(C\xi) = c^2 D(\xi)$
c) $D(C + \xi) = D(\xi)$
d) The sum of the variances of the random variables that are independent of each other, and the sum of their variances, $\sigma^2 = \sigma_1^2 + \sigma_2^2 + \cdots + \sigma_n^2$ is called the additively of variance.

3.1.2 The Basic Concept of Variance and Analysis

First look at a simple example

【Example 3.1.3】 Strength (the same water-cement ratio and slump, to reduce the cement used), the test selected four kinds of dosage, namely the level of $P = 4$, each level repeated test four times, that is, $\gamma = 4$. The strength results at different dosages are listed in Table 3.1.1.

Chapter 3 Variance Analysis of Orthogonal Design

Table 3.1.1 Compressive strength of concrete with different dosage of water reducer (kg/cm^2)

Repeated test times	Strength under different dosages			
	P_1	P_2	P_3	P_4
1	185	173	154	145
2	164	148	140	140
3	192	180	169	158
4	200	174	170	168
Average strength	185.25	168.75	158.25	152.75
Total mean value	166.25			

Form Table 3.1.1:

1. Under the same dosage, the intensity values are different. The reason for this difference is due to a variety of accidental factors in the test process and test methods caused by the difference; this type of error is called the test error.

2. At different dosage, the average of duplicate test data is also different. This is mainly due to changes in the amount of test conditions caused. Due to different dosage caused by the difference in strength is called the conditions worsened.

3. The 16 data in Table 3.1.1 is even more uneven. Their difference is called gross variation. Causes of total variation: First, the test error, and second, the conditions worsened. Analysis of variance the idea of solving this problem is:

a) The experimental errors and conditions are divided by the variation of the data and expressed in terms of the number of them.

b) The condition variation and the test error are compared in a certain sense. The difference between the two people is not obvious, indicating that the change of the condition has little effect on the evaluation index. For example, if there is a big difference among the soldiers and the condition is worse than the test error, Indicating that the changes in conditions on the impact of the superscript is great.

c) Select better process conditions or determine the direction of further testing.

Example 3.1.1 examines only one factor, known as the single factor test. In order to introduce the basic method of ANOVA, we describe the ANOVA method and procedure of Example 3.1.1 single-factor test in detail as follows:

3.1.2.1 Calculate the Total Variation, the Deterioration of the Conditions and the Sum of the Square of the Experimental Error by Degree

1. Total variation I mourning. All the test data in Table 3.1.1 are considered as $x_1, x_2, \cdots, x_n$. and the square sum calculated is S_T, which can be calculated by the above method.

$S_T = 4\ 859$

Its degree of freedom is indicated by f_T, $f_T = pr - 1 = 4 \times 4 - 1 = 15$.

2. Conditions (content) deterioration: with S_A said. It equals the variances of the mean intensities of 185.25, 168.75, 158.25, 152 // 5 for four dosages and the multiplicative squares ($r = 4$) for

each blend. Use the above method to calculate their sum of squares = 613.5, so

$S_A = 4 \times 613.5 = 2\ 454$

Its degree of freedom is expressed in f_A $f_A = p - 1 = 4 - 1 = 3$.

3. Test error: expressed in S_e is the difference between the intensities of doped Pi is the experimental error. The square sum of the test error when doped with:

$S_{e(p_1)} = 185^2 + 164^2 + 192^2 + 200^2 - 0.25 \times (185 + 164 + 192 + 200) = 715$

Similar,

$S_{e(p_2)} = 603$

$S_{e(p_3)} = 605$

$S_{e(p_4)} = 483$

Adding them together is a test error, that is

$S = S_{e(p_1)} + S_{e(p_2)} + S_{e(p_3)} + S_{e(p_4)}$

$= 715 + 603 + 605 + 483$

$= 2\ 406$

From the above results, we find that they are related:

$S_T = S_A + S_e$ \hfill (3.1.7)

$f_T = f_A + f_e$ \hfill (3.1.8)

This means that the sum of squared total variances equals the squared conditional sum plus the squared sum of experimental errors, ie, the additive variability of variances. The variability of total variability equals the sum of the degree of freedom of conditional variation and the degree of experimental error degrees of freedom.

Use Equation 3.1.7 to verify that there is no error in the calculation.

According to the nature of variance (3), $D(C + \xi) = D(\xi)$ for less error in the calculation, the data in Table 3.1.1 minus 140 deducted Kei differential does not pay, Table 3.1.2.

The first row of the lower half of Table 3.1.2; $(\Sigma)^2$ is the sum of the four columns in the same row, the second row CS 5 is the square of the first row, the third row is the square of the four columns in the same column. The sum of the four numbers in each of the three lines is placed on the right-most column and is symbolized. The number of columns is exactly what you want.

$P = \dfrac{1}{n} K^2 = \dfrac{1}{16} \times 420^2 = 11\ 025$

$Q = \dfrac{1}{r} \times 53\ 916 = \dfrac{1}{4} \times 53\ 916 = 13\ 479$

(n is total number of experiment)

$W = 15\ 884$ (the square sum of all data) (r is repeated test times per admixture)

From $P \setminus Q \setminus W$:

$S_A = Q - P = 13\ 479 - 11\ 025 = 2\ 454$

$S_e = W - Q = 15\ 884 - 13\ 479 = 2\ 405$

$S_T = W - P = 15\ 884 - 11\ 025 = 4\ 859$

There is no rounding in the sub-calculations, which is more precision than the previous one. $S_T =$

Chapter 3 Variance Analysis of Orthogonal Design

$S_A + S_e$ Explain the calculation is correct.

Table 3.1.2

Repeated test times r	Strength at all levels ($p=4$) $x_i = R_{28} = 140$				Σ
	P_1	P_2	P_3	P_4	
1	45	33	14	5	
2	2	8	0	0	
3	52	40	29	18	
4	60	34	30	28	
Σ	181	115	73	51	$420 = K$
$(\Sigma)^2$	32 761	13 225	5 329	2 601	$53\,916 = rQ$
Σ^2	8 905	3 909	1 937	1 138	$15\,884 = W$

For the general case, it is to study the effect of a change in the test conditions on the test results. Suppose factor A total of P levels, each level is repeated r times test, the test results are x_{ij}, x_{ij} represents the j-test under the conditions of the pull. The schedule shown in Table 3.1.3 is tabulated.

Table 3.1.3

	A_1	A_2	...	A_i	...	A_p	Σ
1	x_{11}	x_{21}	...	x_{i1}	...	x_{p1}	
2	x_{12}	x_{22}	...	x_{i2}	...	x_{p2}	
$\vdots$	$\vdots$	$\vdots$	...	$\vdots$	...	$\vdots$	
j	x_{1j}	x_{2j}	...	x_{ij}	...	x_{pj}	
$\vdots$	$\vdots$	$\vdots$	...	$\vdots$	...	$\vdots$	
r	x_{rj}	x_{2r}	...	x_{tr}	...	x_{pr}	
Σ	$\sum_{j=1}^{r} x_{1j}$	$\sum_{j=1}^{r} x_{2j}$	...	$\sum_{j=1}^{r} x_{ij}$	...	$\sum_{j=1}^{r} x_{pj}$	$\sum_{i=1}^{p}\sum_{j=1}^{r} x_{ij} = K$
$(\Sigma)^2$	$(\sum_{j=1}^{r} x_{1j})^2$	$(\sum_{j=1}^{r} x_{2j})^2$	...	$(\sum_{j=1}^{r} x_{ij})^2$	...	$(\sum_{j=1}^{r} x_{pj})^2$	$\sum_{i=1}^{p}(\sum_{j=1}^{r} x_{ij})^2 = rQ$
Σ^2	$\sum_{j=1}^{r} x_{1j}^2$	$\sum_{j=1}^{r} x_{2j}^2$	...	$\sum_{j=1}^{r} x_{ij}^2$	...	$\sum_{j=1}^{r} x_{pj}^2$	$\sum_{i=1}^{p}\sum_{j=1}^{r} x_{ij}^2 = W$

$$P = \frac{1}{pr}\left(\sum_{i=1}^{p}\sum_{j=1}^{r} x_{ij}\right)^2$$

$$Q = \frac{1}{pr}\sum_{i=1}^{p}\left(\sum_{j=1}^{r} x_{ij}\right)^2$$

$$W = \sum_{i=1}^{p}\sum_{j=1}^{r} x_{ij}^2$$

For write simply,

$$K_i = \sum_{j=1}^{r} x_{ij} \qquad K = \sum_{i=1}^{p}\sum_{j=1}^{r} x_{ij}^2$$

$$x_i = \frac{1}{r}\sum_{j=1}^{r} x_{ij} = \frac{1}{r}K_i$$

It is:

$$P = \frac{1}{n}K^2, Q = \frac{1}{r}\sum_{i=1}^{p} K_i \quad n = pr$$

$$\overline{X} = \frac{1}{pr}K, \overline{X} = \frac{1}{n}K$$

From the calculation of Example 3.2.1, we see:

$$S_A = r\sum_{i=1}^{p}(X_i - \overline{X})^2$$

$$S_e = \sum_{i=1}^{p}\sum_{j=1}^{r}(x_{ij} - \overline{X}_i)^2$$

$$S_T = \sum_{i=1}^{p}\sum_{j=1}^{r}(x_{ij} - \overline{X})^2$$

Computation of these three formulas is not convenient, similar to formula 3.1.4 can be introduced:

$$S_A = Q - P \tag{3.1.9}$$
$$S_e = W - Q \tag{3.1.10}$$
$$S_T = W - P \tag{3.1.11}$$

Their corresponding degrees of freedom are:

$$f_A = P - 1, f_e = (r-1), f_T = pr - 1 \tag{3.1.12}$$

3.1.2.2 Calculate the Mean Square, According to the Definition, from Formula 3.1.5:

$$f_A = S_A/f_A = 2\,454/3 = 818$$
$$S_e = S_e/f_e = 2\,405/12 = 200.5$$

3.1.2.3 Factor Significance Test

The so-called factor significance test is to determine whether the impact of factors on the indicators when the level of change is significant. It has been pointed out earlier that the mean square reflects a measure of the magnitude of the fluctuations and that the magnitude of the comparison show that the influence of the content on the compressive strength is significant. How to judge? Put statistics

$$F = \frac{\overline{S_A}}{\overline{S_e}} \tag{3.1.13}$$

This test is called the f test. Appendix m table two were given four cases of heart straight, Lang significant level, $a = 0.20$, $a = 0.10$, $a = 0.05$, $a = 0.01$. There are two parameters on the table, and f_1, f_2, $F_a(f_1, f_2)$ that significant level of a, the corresponding two degrees of freedom for the ruler and time table values. For example $F_{0.05}(3,12) = 3.5$, $F_{0.01}(3,12) = 6.0$.

The value on the F table is the critical value used to determine if the effect of the factor is significant. For single factor test, take $f_1 = F_a$, $f_2 = f_e$. Comparing the P value calculated by formula 3.1.13 with the value on the table, there are four cases:

1. $F > F_{0.01}$, said that the impact of dosage is particularly significant, recorded as " * * ";
2. $F_{0.01} \geq F > F_{0.05}$, said the amount of significant impact, denoted as " * * ";

3. $F_{0.05} \geq F > F_{0.10}$ means that the amount of a certain impact, recorded as " (*) "," (•) ";
4. $F_{0.10} \geq F$, That can not see the amount of the greater impact on the strength for this case, $F = 818/200.5 = 4.08$, because

$$F_{0.05}(3,12) = 3.50 < F = 4.08 < F_{0.01}(3,12) = 6.0$$

Therefore, the effect of dosage on strength is significant.
The results are listed in ANOVA Table 3.1.4.

Table 3.1.4 Variance analysis table

Variation source	Sum of squares	Freedom	Mean square	F value	Critical value
Volume	S_A = 2 454	3	818	4.08*	$F_{0.05}(3,12) = 3.5$
Error	S_C = 2 450	12	200.4		$F_{0.01}(3,12) = 6.0$
SUM	S_T = 4 859	15			

From the results of analysis of variance, since the dosage has a significant impact on the strength, so the process conditions should generally choose the most corresponding strength of the #5, that is, choose P_1 volume.

Finally need to explain two points:

1. Is there any significant effect of the dosage on the strength? It is concluded through 16 tests. Due to the experimental error, it is very hard for us to have 100% conclusion. The significance level of $a = 0.05$, indicating the following conclusion: "a mixed with a significant impact on the degree of difficulty," 95 times in 100 times is correct, the She 5 times can make mistakes.

2. When f_e the hole is very small, the sensitivity of the F test is very low, that is, the factor of the North Korea has a significant impact, but the F test can not be determined. The f_e larger, the higher the test sensitivity: Too large to require more tests will increase, but this is not cost-effective, this is a contradiction. In the general case of hope, hope

In the range of 5 – 10, as it is actually difficult, we will relax 0.20 when doing F-test and pay attention to the conclusion made in further tests in practice.

3.2 Variance Analysis of Orthogonal Design

3.2.1 The Basic Methods and Characteristics

Anova method is used for data analysis in orthogonal design. In order to facilitate the relationship between Anova and Anova, Anova is carried out in the following with reference to the strength test result of Example 1.2.1.

【Example 3.2.1】 The visual analysis results of this example are shown in Table 3.2.1.

According to Equation 3.1.7 and 3.1.8, the square degree of freedom in this example is decomposed into:

$$S_T = S_A + S_B + S_O + S_E \tag{3.2.1}$$

$$f_T = f_A + f_B + f_O + f_E \tag{3.2.2}$$

Let's give a general formula for calculating them. *Let orthogonal design* made a total of n times test,

test results for x_1, x_2, $\cdots$, x_n, r_a, r_b, r_c respectively table. In this example, $n = 9$, $r_a = r_b = r_c = 3$. Similar to single factor analysis of variance. Show three factors A, B, C for each level of trial repetition. This example

Table 3.2.1 Visual analysis results

Experimental number	A. Slag	B. Gypsum	C. Iron powder		28-day Compressive strength (kg/cm^2)		
	1	2	3	4	R_{28}	$X'_i = R_{28} - 830$	X'^2_i
1	1(10%)	1(2%)	1(3%)	1	765	-65	4 225
2	1(10%)	2(3.5%)	2(6%)	2	810	-20	400
3	1(10%)	3(5%)	3(9%)	3	758	-72	5 184
4	2(15%)	1(2%)	2(6%)	3	857	27	729
5	2(15%)	2(3.5%)	3(9%)	1	891	61	3 721
6	2(15%)	3(5%)	1(3%)	2	765	-65	4 225
7	3(20%)	1(2%)	3(9%)	2	907	77	5 929
8	3(20%)	2(3.5%)	1(3%)	1	867	37	1 869
9	3(20%)	3(5%)	2(6%)	1	860	30	900
K_1	-157	39	-93	26	$\Sigma = 10$		$\Sigma = 26\ 682$
K_2	23	78	37	-8	$K = 10$		
K_3	144	-107	66	-8			
R	301	185	159	34			

$$\left. \begin{aligned} K &- x_1 \mid x_2 \mid \cdots, + x_n \\ P &= \frac{1}{n} \cdot K^2 \\ W &= \sum_{i=1}^{n} x_i^2 \\ Q_A &= \frac{1}{r_n} \sum_{i=1}^{r_a} (K_i^A)^2 \\ Q_B &= \frac{1}{r_n} \sum_{i=1}^{r_b} (K_i^B)^2 \\ Q_C &= \frac{1}{r_n} \sum_{i=1}^{r_c} (K_i^C)^2 \end{aligned} \right\} \quad (3.2.3)$$

Their sum of squares is:

$$\left. \begin{aligned} S_A &= Q_A - P \\ S_B &= Q_B - P \\ S_C &= Q_C - P \\ S_T &= Q_T - P \end{aligned} \right\} \quad (3.2.4)$$

Direct calculation & sometimes more trouble, usually according to Equation 3.2.1 meter, that is,

$$S_e = S_T - S_A - S_B - S_C$$

The total degree of freedom f_T equals the number of trials n minus 1, ie $f_T = n - 1 = 9 - 1 = 8$, The degree of freedom of each factor is equal to the number of levels minus 1, 2, $f_A = f_B = f_C = 3 - 1 = 2$

Chapter 3 Variance Analysis of Orthogonal Design

Using the above formula to calculate Table 3.2.1, the data in the table are subtracted by 830, which is not affected by the sum of squares calculation. Substitute the data into Equation 3.2.3, Each calculated:

$$K = \sum_{i=1}^{n} x'_i = -65 - 20 + \cdots + 30 = 10$$

$$P = \frac{1}{n} \cdot K^2 = \frac{1}{9} \times 10^2 = 11$$

$$W = \sum_{i=1}^{n} x'^2_i = 26\ 682$$

$$Q_A = \frac{1}{r_a} \sum_{i=1}^{r_a} (K_i^A)^2 = \frac{1}{3} \times [(-157)^2 + 23^2 + 144^2] = 15\ 305$$

$$Q_B = \frac{1}{r_b} \sum_{i=1}^{r_b} (K_i^B)^2 = \frac{1}{3} \times [(39)^2 + 78^2 + (-107)^2] = 6\ 351$$

$$Q_C = \frac{1}{r_a} \sum_{i=1}^{r_c} (K_i^C)^2 = \frac{1}{3} \times [(-93)^2 + 37^2 + 66^2] = 4\ 791$$

From Equation 3.2.4, their sum of squares is:

$S_A = Q_A - P = 15\ 305 - 11 = 15\ 294$
$S_B = Q_B - P = 6\ 531 - 11 = 6\ 340$
$S_C = Q_C - P = 4\ 709 - 11 = 4\ 780$
$S_T = W - P = 26\ 682 - 11 = 26\ 671$

And $S_e = S_T - S_A - S_B - S_C = 26\ 671 - 15\ 294 - 6\ 340 - 4\ 780 = 257$, according to Equation 3.2.2, so:

$$f_e = f_T - f_A - f_B - f_C = 8 - 2 - 2 - 2 = 2$$

The results of the calculation of variance analysis Table 3.2.2.

Table 3.2.2 Variance analysis table

Variation source	Sum of squares	Freedom	Mean square	F value	Critical value
Slag Volume	A. $S_A = 15\ 294$	2	7 647	29.5	$F_{0.01}(2,2) = 99.0$
Gypsum	B. $S_B = 6\ 340$	2	3 170	24.7	$F_{0.03}(2,2) = 19.0$
Iron powder	C. $S_C = 4\ 780$	2	2 390	18.6	$F_{0.1}(2,2) = 9.0$
Error	D. $S_e = 257$	2	128.5		
SUM	E. $S_T = 26\ 671$	8			

From the analysis of variance results:

1. The order of the major and minor intensities of the various shadows is: slag content Shi Dian doped iron content, which is consistent with the results of Table 1.2.2.
2. The slag pick-up basket and stone-grained incorporation had a significant effect on the intensity of 28 d, while iron-bound fish had a certain impact on the strength.
3. The experimental error in this example is $\sqrt{128.5} = 11.3$ kg/cm^2. Very close to the intuitive analysis of empty columns.

After done a significant test, then choose the best process conditions or cooperation.

Analysis of variance analysis point of view: just select the obvious factors on the line, insignificant factors, in principle, optional within the scope of any one test. In this case, the first two due to the music are significantly due to the optimal level of each factor to select the maximum value of K corresponding to the level, that is A_3 and B_2; iron powder mixing intensity of certain funeral, in principle, the three levels Select, now take C_3. Therefore, the optimal formula is if we further analyze factor $A_3B_2C_3$ (multiple comparisons, discussed later), we find that there is no significant difference between B_2 and B_1, so the optimal formulation is chosen as $A_3B_1C_3$ the combination of the conditions of the No. 7 test.

In this example, we can look at the variance analysis of orthogonal designs with the following features:
1. The sum of squares is equal to the sum of squares of the columns. In this case, there are no factors in the fourth column. If we follow the calculation of the sum of squares of each factor and calculate the value of K in column 4 (see Table 3.2.1)

$$Q_4 = \frac{1}{3} \times [26^2 + (-8)^2 + (-8)^2] = 268$$

$$S_e = Q_4 - P = 268 - 11 = 257$$

Exactly ten equal to S_e, $L_9(3^4)$ orthogonal table of four square sum together, exactly equal to the total sum of squares. The advantage of ANOVA is to factorize the sum of squares into the sum of the squares of the factors and the error. Orthogonal design fixes this decomposition to each column, what is given to a column when it is scheduled, and the square sum of the column reflects that. The square sum of the column reflects the experimental error for an empty column with no scheduling factor. Therefore, S_e is not necessarily calculated by Equation 3.2.1, but can be directly calculated by the column without arrangement factor.

2. Calculation is normalized. The calculation steps for each factor in orthogonal design are exactly the same, and each factor corresponds to a certain column. If a column is an error, the corresponding square sum is exactly the same as the factor column. This is easy to remember, but also conducive to the preparation of electronic computer programs.

From the two properties of sub-analysis, the basic calculation of variance analysis can be applied to each column. Suppose there are P levels in a column and r tests at the same level. K_1, K_2, $\cdots$, k_p represent the sum of r data of the corresponding p levels, so:

$$K = K_1 + K_2 + \cdots + K_p$$

$$P = \frac{1}{pr} \cdot K^2$$

$$Q = \frac{1}{r} \sum_{i=1}^{p} K_i^2$$

Thus, the square sum of squares with the corresponding degree of freedom is:

$$S = Q - P$$

$$f = p - 1$$

3. Easy to analyze the factors of the primary and secondary. The primary and secondary discriminant principle is to compare their mean square, square is the major factors, and small is the secondary factor. In this example, according to the size of the mean square, the order of the primary and secondary factors that affect the strength of the discriminant is: slag mixed→with a stone doped→ iron play mixed.

Chapter 3 Variance Analysis of Orthogonal Design

3.2.2 Several Common Analysis of Variance

3.2.2.1 Blank out more than One Column of Variance Analysis

【Example 3.2.2】 The quality of naphthalene-type DH super plasticizer was tested on three different technological products: DH - 1, DH - 2 and DH - 3, Concrete test (water-cement ratio is fixed at 0.6, the depression between 5.7 - 8.2 cm), Which is 0.3%, 0.5%, 0.7%. The 28 d compressive strength data obtained from the test were analyzed by ANOVA.

The factors and levels in the test are listed in Table 3.2.3.

Table 3.2.3 Factor level table

Factor	Level		
	1	2	3
A. Volume(%)	0.3	0.5	0.7
B. Squeeze	DH - 1	DH - 2	DH - 3

Assessment indicators: 28 d compressive strength.

$L_9(3^4)$ Test results and calculation results are shown in Table 3.2.4.

In this case $n = 9$, the number of columns in each column $P = 3$, the number of tests with the level of $r = 3$. Substituting the data into Equation 3.2.3 calculates:

Table 3.2.4 $L_9(3^4)$ Test result of experiment scheme

Experimental number	A. Volume	B. Squeeze			28-day Compressive strength X_i (kg/cm^2)
	1	2	3	4	
1	1(0.3)	1(DH - 1)	3	2	182
2	2(0.5)	1(DH - 1)	1	1	158
3	3(0.7)	1(DH - 1)	2	3	163
4	1(0.3)	2(DH - 2)	3	3	191
5	2(0.5)	2(DH - 2)	3	3	187
6	3(0.7)	2(DH - 2)	1	2	171
7	1(0.3)	3(DH - 3)	1	3	200
8	2(0.5)	3(DH - 3)	2	2	200
9	3(0.7)	3(DH - 3)	3	1	184
K_1	573	503	529	533	
K_2	545	549	554	553	SUM: $K = 1\ 636$
K_3	518	584	553	550	
R	55	81	25	20	

$$P = \frac{1}{n} \cdot K^2 = \frac{1}{9} \times 1\,636^2 = 297\,388$$

$$W = \sum_{i=1}^{n} X_i^2 = 182^2 + 158^2 + \cdots + 184^2 = 299\,204$$

$$Q_A = \frac{1}{r}\sum_{i=1}^{r}(K_i^A)^2 = 0.333 \times (573^2 + 545^2 + 518^2) = 297\,893$$

$$Q_B = \frac{1}{r}\sum_{i=1}^{r}(K_i^B)^2 = 0.333 \times (503^2 + 549^2 + 584^2) = 298\,489$$

$$Q_3 = \frac{1}{r}\sum_{i=1}^{r}(K_i^3)^2 = 0.333 \times (529^2 + 554^2 + 553^2) = 297\,522$$

$$Q_4 = \frac{1}{r}\sum_{i=1}^{r}(K_i^4)^2 = 0.333 \times (533^2 + 553^2 + 550^2) = 297\,466$$

$S_A = Q_A - P = 297\,893 - 297\,388 = 505$

$f_A = p - 1 = 3 - 1 = 2$

$S_B = Q_B - P = 297\,522 - 297\,388 = 1\,101$

$f_A = p - 1 = 2$

$S_1 = Q_1 - P = 297\,466 - 297\,388 = 134$

$f_1 = p - 1 = 3 - 1 = 2$

$S_4 = Q_4 - P = 297\,466 - 297\,388 = 78$

$f_4 = p - 1 = 3 - 1 = 2$

$S_T = W - P = 299\,204 - 297\,388 = 1\,816$

$f_T = pr - 1 = 3 \times 3 - 1 = 8$

$S_e = S_T - S_A - S_B = 1\,816 - 505 - 1\,101 = 210$

$f_e = (p-1)(r-1) = 2 \times 2 = 4$

$S_e = S_3 + S_4 = 134 + 78 = 212$

$f_e = f_3 + f_4 = 2 + 2 = 4$

Can be seen, empty column 3, 4 showed the test error.

ANOVA is listed in Mourning Table 3.2.5.

Table 3.2.5 Variance analysis table

Variation source	Sum of squares	Freedom	Mean square	F value	Critical value
Volume	A. $S_A = 505$	2	252.5	4.81	$F_{0.01}(2,4) = 18.0$
Squeeze	B. $S_B = 1\,101$	2	550.5	10.49	$F_{0.05}(2,4) = 6.9$
Error	C. $S_e = 210$	4	52.5		$F_{0.1}(2,4) = 4.3$
SUM	D. $S_T = 1\,816$	8			

From the analysis of variance results:

a. The super plasticizer has a significant effect on the strength, DH-3 is the best, DH-2 is the

Chapter 3 Variance Analysis of Orthogonal Design

second, DH-1 is the worst.

b. The dosage of super plasticizer also has a certain influence on the strength, with the increase of doping and the decrease of strength.

c. The test in this case is wrongly measured as $\sqrt{S_e} = \sqrt{52.5} = 7.2 \text{ kg/cm}^2$.

Careful readers may have found that this case is an analysis of the variance of a two-factor crossover comprehensive trial. Although the number of trials is not a funeral, we consciously arranged this way so that analysis of the problem clear conclusion, law ridiculous. It also shows that in the absence of interaction between factors, the empty columns (regardless of a few columns) on the orthogonal table all appear as experimental errors.

3.2.2.2 Repeated Trial Analysis of Variance

As mentioned earlier, empty columns are very poorly used to analyze test failures. However, when the number of columns of the orthogonal table is full, the degree of freedom for which the error must be provided from the repeated trial data or when the degree of freedom (error) is small is low, that is, the test sensitivity is low, that is, Poor have a significant impact, but f test can not judge, then also need to make repeated tests. The so-called repeat test is to test the same conditions repeated one or more times. Under the same conditions, the difference between several test results reflects the experimental error.

【Example 3.2.3】 The effects of water-cement ratio, sand fineness modulus and tamping method cement mortar compressive strength was investigated to select the mixing ratio and forming method of 800 high-strength cement mortar.

Other test conditions: Jingmen 525 Portland cement, river sand. Daphnia was fixed at 238 kg/m³, water reducer FDN content of 0.5%, mortar calculated capacity of 2 260 kg/m³.

The factor and horizontal are shown in Table 3.2.6.

Table 3.2.6 Factor level table

Level	Factor		
	A. Water cement ratio	B. Sand fineness magic	C. Tamping method
1	0.26	3.22	Intercalation
2	0.29	2.84	Vibration 15 s
3	0.32	1.96	Vibration 30 s

Assessment indicators: 28-day compressive strength.

The test plan is shown in Table 3.2.7. Each test number repeated test twice, test results and its K_i calculation are also listed in the table.

This is an orthogonal design with repeated trials, the calculation of which is slightly different from Example 3.2.1. Let the x_{ij} test, which represents the i trial number.

Table 3.2.7 Test scheme and the result of extreme difference calculation

Experimental number	A	B	C		28-day Compressive strength(kg/cm^2)		SUM: X_i
	1	2	3	4	1	2	
1	1(0.26)	1(3.22)	1(intercalation)	1	840	814	1 654
2	1	2(2.84)	2(vibration 15 s)	2	793	779	1 572
3	1	3(1.96)	3(vibration 30 s)	3	770	771	1 541
4	2(0.29)	1	2	3	789	809	1 598
5	2	2	3	1	800	792	1 592
6	2	3	1	2	737	755	1 492
7	3(0.32)	1	3	2	800	802	1 602
8	3	2	1	1	729	759	1 488
9	3	3	2	1	709	702	1 411
K_1	4 746	4 854	4 634	4 657			
K_2	4 582	4 652	4 534	4 666		Σ = 13 950	
K_3	4 501	4 444	4 735	4 627			
R	266	410	154	39			

The test number is repeated r times, so that there are $X_i = \sum_{j=1}^{r} x_{ij}$, $x_i = \frac{1}{r}X_i$, n test numbers. Obviously, the difference between the r data of the same test is the test error, as in the previous case,

$$S_e = \sum_{i=1}^{n} \sum_{j=1}^{r} (x_{ij} - x_i)^2$$

Let the average of all data, ie

$$\bar{x} = \frac{1}{nr} \sum_{i=1}^{n} \sum_{j=1}^{r} x_{ij}^2$$

Then the total square and Sr is the overall data of the deterioration

$$S_T = \sum_{i=1}^{n} \sum_{j=1}^{r} (x_{ij} - \bar{x})^2$$

Suppose there is p level in a certain column of orthogonal table, and there are q test numbers (apparently $qp = n$) for each level. The mean value of p levels in this column is: $k_1, k_2, \cdots K_p$, then the square sum of the column is:

$$S = qr \sum_{i=1}^{q} (\bar{K}_i - \bar{x})^2$$

These formulas are very inconvenient to calculate, the practical use of the following formula:

$$K = \sum_{i=1}^{n} \sum_{j=1}^{r} x_{ij}$$

$$P = \frac{1}{nr} K^2$$

$$W = \sum_{i=1}^{n} \sum_{j=1}^{r} x_{ij}^2 \quad\quad (3.2.5)$$

$$R = \frac{1}{r} \sum_{i=1}^{n} x_i^2$$

which is,

$$S_e = W - R$$
$$S_T = W - P \quad\quad (3.2.6)$$

Suppose a column has p levels, each level has q test numbers, and the sum of the horizontal test data is the ruler $K_1, K_2, \cdots, K_p$, then the square sum of the columns is:

$$S = Q - P$$

among them

$$Q = \frac{1}{qr} \sum_{i=1}^{p} (K_i)^2 \quad\quad (3.2.7)$$

for this example:

$n = 9, r = 2$

$K = 840 + 793 + \cdots + 702 = 13\ 950$

$P = \dfrac{13\ 950^2}{9 \times 2} = 10\ 811\ 250$

$W = 840^2 + 793^2 + \cdots + 702^2 = 10\ 834\ 898$

$R = 0.5 \times (1\ 654^2 + 1\ 572^2 + \cdots + 1\ 411^2) = 10\ 833\ 591$

so,

$S_e = W - P = 10\ 834\ 898 - 10\ 833\ 591 = 1\ 307$

$S_T = W - P = 10\ 834\ 898 - 10\ 811\ 250 = 23\ 648$

For factors 1, 2 and 3 where factors A, B and C are located, $p = 3$, $q = 3$ and $r = 2$, then

$$Q_A = \frac{1}{3 \times 2} \times (4\ 767^2 + 4\ 782^2 + 4\ 501^2) = 10\ 817\ 402.33$$

$$S_A = Q_A - P = 10\ 817\ 402.33 - 10\ 811\ 250 = 6\ 152.33$$

$$Q_B = \frac{1}{3 \times 2} \times (4\ 854^2 + 4\ 531^2 + 4\ 444^2) = 10\ 825\ 259.33$$

$$S_B = Q_B - P = 1\ 081\ 602.33 - 13\ 411\ 250 = 14\ 009.33$$

$$Q_C = \frac{1}{3 \times 2} \times (4\ 654^2 + 4\ 581^2 + 4\ 735^2) = 10\ 813\ 290.33$$

$$S_C = Q_C - P = 1\ 091\ 602.33 - 10\ 811\ 250 = 2\ 050.33$$

According to the general rule of analysis of variance should be

$$S_T = S_A + S_B + S_C + S_e$$

but here,

Orthogonal Design in Concrete Application

$S_A + S_B + S_C + S_e = 6\ 152.33 + 14\ 009.33 + 2\ 042.33 + 1\ 307 = 23\ 509 < S_T = 23\ 648$

Why is that? Because we only used the first three colunms of data, ond one column didn't make use of it. This coloum can be used to estimnte errors if there are no repeareted tvials and no interactions be tweens factors; If trere are intereations It can't be used to estimate errors, Generally speaking. the sum of squares in this column is taken as S_e, and the $S_e = WR$; S_e calcalute is used to carry out F test for S_e, If A and B can, owe similar, the interaction between factors can't he seen, then A and B can ecombimed to estimate the error. If A is much larger than B, F test is significant. indicating that the interaction betweenfacto s can not be ignored.

Calculate S_{e1}, there are two ways, one from $L_9(3^4)$ direct calculation, which is

$Q_4 = \dfrac{1}{3 \times 2} \times (4\ 657^2 + 4\ 666^2 + 4\ 627^2) = 10\ 811\ 389$

$S_4 = Q_4 - P = 10\ 811\ 389 - 10\ 811\ 250 = 139$

the other way is:

$S_{e1} = R - P - S_A - S_B - S_C = 10\ 833\ 591 - 10\ 811\ 250 - 6\ 252.33 - 14\ 009.33 - 2\ 040.33 = 139$

Explain the calculation is correct. Two methods depending on the calculation to facilitate the choice of use. The calculation results are listed in Anova Table 3.2.8. Because S_{e1} and S_e there is no significant difference, they combined together to estimate the error.

Table 3.2.8 Variance analysis table

Variation source	Sum of squares	Freedom	Mean square	F value	Critical value
A	$S_A = 6\ 152.33$	2	3 076.2	23.4	$F_{0.01}(2,11) = 7.2$
B	$S_B = 14\ 009.33$	2	7 004.7	53.3	
C	$S_C = 2\ 040.33$	2	1 020.2	7.76	
Empty column	$S_{e1} = 139$	2	69.5 }$\overline{S}_e = 131.4$ 145.2		
Error	$S_e = 1\ 307$	9			
SUM	$S_T = 23\ 648$	17			

Analysis of variance showed that:

1. From the mean square size, it can be seen that the major and minor order of the influencing factors of the factors are sand fineness modulus (B) →water − cement ratio (A) →transaction mode (C);
2. The various factors have a particularly significant effect on the 28 d pit pressure of the mortar. (it is meet design strength condition) The combined condition required to achieve the design strength of the highest strength is $A_1 B_1 C_3$, ie 0.26 for water-cement coarse sand and 30 s for coarse sand.
3. Since S_{e1} can be regarded as an error, we can not see the interaction between the factors on the intensity.
4. The experimental error in this case is $\sqrt{\overline{S}_e} = \sqrt{131.4} = 11.5$ kg/cm^2, that the test in Kyoto is quite high.

Chapter 3 Variance Analysis of Orthogonal Design

3.2.2.3 Repeated Sampling Analysis of Variance

The so-called repeated sampling, is obtained in the same test at the same time three test blocks for strength test, the difference between the strength of three test blocks, reflecting the repeated sampling error (also known as intra – group test error). Repeated sampling under certain conditions can also be used to estimate the experimental error, but the error between the test blocks is only partial, less than the error of repeated experiments. In order to save duplication of effort, the data provided by repeated sampling can be used to test the significance of the factors. According to experience, when testing, if about half of the factors are not significant because of the conflict between the primary and secondary, you can think of this test is reasonable. Otherwise, silkworm complex test is still as Velvet.

【Example 3.2.4】 In this example, we introduce the method of daddy analysis with repeated sampling. The test program in Table 2.1.2 copied over, each test number repeated three-test blocks, the test results for a simple calculation tabulated in Table 3.2.9.

As in Example 3.2.3, calculate the total sum of squares, the sum of squares of the columns, the square of the error, and their degrees of freedom. The calculation results are listed in ANOVA Table 3.2.10. Since there are no significant differences between S_{e1}, S_D and S_e, they are combined to calculate the error.

The results of variance analysis show that:
1. The influence of lime-water ratio and vibrating conditions on the strength is particularly significant, and the content of coal ash has a certain influence on the intensity, while the influence of oil-washing blending is within the error range.
2. The vibration of mobile concrete can effectively improve its strength.
3. Test error $\sigma_e = \sqrt{\bar{S}_e} = \sqrt{534.616} = 23.1 \text{ kg/cm}^2$.

Table 3.2.9 Test results and calculation

Experimental number	A	B	C	D		28-day Compressive strength (kg/cm²)			SUM: X_t
	1	2	3	4	5	1	2	3	
1	1(0)	1(2.0)	2(5)	2(0.7)	1	289	345	336	940
2	3(30s)	2(2.4)	2	1(0.6)	1	486	519	491	1 499
3	2(15s)	2	2	2	2	511	456	477	1 444
4	4(0s)	1	2	1	1	456	431	410	1 297
5	1	2	1(0)	1	1	401	443	393	1 237
6	3	1	1	2	2	456	452	426	1 334
7	2	1	1	1	1	431	367	435	1 233
8	4	2	1	2	2	515	532	536	1 533
$K_1(\bar{K}_1)$	2 177(363)	4 804(406)	5 387(449)	5 226(439)	5 255(438)				
$K_2(\bar{K}_2)$	2 677(446)	5 763(480)	5 180(432)	5 301(442)	5 312(443)				
$K_3(\bar{K}_3)$	2 833(472)						Σ = 10 567		
$K_4(\bar{K}_4)$	2 880(480)								
R	2 880(117)	859(74)	207(17)	35(3)	57(5)				

Table 3.2.10 Variance analysis table

Variation source	Sum of squares	Freedom	Mean square	F value	Critical value
A	$S_A = 51\ 762.458$	3	17 254.153	32.27	$F_{0.01}(3,18) = 5.1$
B	$S_B = 38\ 820.041$	1	38 820.041	71.68	$F_{0.01}(1,18) = 8.3$
C	$S_C = 1\ 758.374$	1	1 785.374	3.44	$F_{0.05}(1,18) = 4.4$
D	$S_D = 41.041$	1	$\left.\begin{array}{l}51.041\\135.377\\589.792\end{array}\right\} \bar{S}_e = 534.616$		$F_{0.1}(1,18) = 3.0$
Empty column	$S_E = 135\ 377$	1			
Error	$S_e = 9\ 436.667$	16			
SUM	$S_T = 101\ 490.958$	23			

Discussion:

1. As we have pointed out in Example 2.1.1, for the mixed level table, we should judge the order of the primary and secondary factors according to the index and the range difference $\bar{R}$, but not the range difference $\bar{R}$ of the index average. A→B→C→D As you can see from the data in the last row of Table 3.2.9, if you press R, the order of major and minor is: A→B→C→D, which is obviously the wrong conclusion.

2. $L_8(4^1 \times 2^4)$ Four factors are arranged on the orthogonal table If there is no re-test data available, the total degree of freedom is 7, the degree of error freedom for a long time $f_e = 1$, and when it is small, the sensitivity of the F-test is very low. Due to the use of complex sample data, the accuracy of F greatly improved.

3.2.2.4 The Proposed Level of Analysis of Variance

【Example 3.2.5】 Minjiang Project Bureau of Fujian Province conducted a wet-fly ash process test at Tan Hydropower Station, and wet-mixed fly ash reached more than 1,000 t in this project. In the experiment, orthogonal design was adopted to optimize the process parameters. We conducted further analysis of variance and discussion based on the data obtained from the experiments.

Assessment indicators: 28 d compressive strength, strength, the more the better.

1. The factors and levels are shown in Table 3.2.11. The third level of turpentine soap D adulteration due to is to be level. The water-cement ratio in the table A is the ratio of water to fly ash weight.

Table 3.2.11 Factor level table

Factor	Level					
	1	2	3	4	5	6
A. Water cement ratio	0.65	0.70	0.75	0.80	0.85	0.90
B. Mixing time (min)	5	10	15			
C. Wind pipe root number	2	3	4			
D. Dosage of rosin soap (1/10 000)	2.2	0	2.2			
E. Cone angle (°)	120	90	60			
F. Mixing time	1.0	1.5	2.0			

Other conditions in the test: ordinary cement Yong'an 500; fly ash mixed is 30%, gravel, three graded, concrete water-cement ratio weight ratio of water to cement) is 0.65.

Chapter 3 Variance Analysis of Orthogonal Design

2. Test program and the calculation results are shown in Table 3.2.12.
3. The analysis of test results

a) Straight to the analysis

Directly to see the strength of the first 4 test 300.1 kg/cm² combination of living conditions, $A_2 B_2 C_1 D_2 E_3 F_1$.

Calculate the factors that affect the strength of yesterday's order was: $E \rightarrow C \rightarrow A \rightarrow B \rightarrow F \rightarrow D$, that is, the bottom cone angle E is the most important factor affecting strength, the second main factor C is duct root The number of water-cement than the impact of A ranks third, and the impact of other factors are small. Get the combination of higher intensity conditions $A_4 B_2 C_1 D_1 E_3 F_2$ heart.

In both combinations, the best levels of C and E that play a major role are the same, $C_1 E_3$. The selection of other factors, the need for further analysis of variance after the decision.

b) Variance analysis

The sum of squares and their degrees of freedom are as follows:

Table 3.2.12 $L_{18}(6 \times 3^6)$ Test results and calculation

Experimental number	A	B	C	D	E	F	R_{28}
	1	2	3	4	5	6	7	(kg/cm²)
1	1 (0.65)	1	3 (15)	2 (3)	2 (0)	1 (120)	2 (1.5)	160.4
2	1	2	1 (5)	1 (2)	1 (2.2)	2 (90)	1 (1.0)	216.1
3	1	3	2 (10)	3 (4)	3 (2.2)	3 (60)	3 (2.0)	121.7
4	2 (0.75)	1	2	1	2	3	1	300.1
5	2	2	3	3	1	1	3	133.4
6	2	3	1	2	3	2	2	198.1
7	3 (0.80)	1	1	3	1	3	2	185.6
8	3	2	2	2	3	1	1	114.6
9	3	3	3	1	2	2	3	145.6
10	4 (0.85)	1	1		3	1	3	191.2
11	4	2	2	1	2	2	2	136.3
12	4	3	3	3	1	3	1	270.9
13	5	1	3	2	3	2	1	105.5
14	5	2	1	3	2	3	3	219.9
15	5	3	2	2	1	1	2	167.5
16	6 (0.90)	1	2	1	1	2	3	240.6
17	6	2	3	2	3	3	2	298.4
18	6	3	1	1	2	1	1	64.2
K_1	589.2	1 183.1	1 047.8	1 318.9	1 213.8	831.3	10 714	
K_2	631.6	1 218.7	1 271.8	1 204.5	1 126.5	1 142.2	1 246	
K_3	455.5	1 059	1 114.2	937.4	1 150.5	1 487.3	1 143.4	
K_4	689.4							SUM:3 460.8
K_5	492.9							
K_6	603.2							
R	252.9	159.7	197.0	381.5	93.3	656	174.6	

Orthogonal Design in Concrete Application

$K = 3\ 460.8$

$P = \dfrac{1}{18} \times 3\ 460.8^2$

$W = 737\ 813.5$

$S_T = W - P = 737\ 813.5 - 665\ 396.5 = 72\ 416, f_T = 17$

$Q_A = \dfrac{1}{3} \times (589.2^2 + \cdots + 603.2^2) = 679\ 702.9$

$S_A = Q_A - P = 679\ 702.9 - 665\ 396.5 = 14\ 306.4$

$f_A = 6 - 1 = 5$

$Q_B = \dfrac{1}{6} \times (1\ 074.8^2 + \cdots + 1\ 114.2^2) = 669\ 018.7$

$S_B = Q_B - P = 3\ 622.2$

$f_B = = 3 - 1 = 2$

$Q_C = \dfrac{1}{6} \times (1\ 318.8^2 + \cdots + 937.4^2) = 678\ 172.7$

$S_C = Q_C - P = 12\ 776.2$

$f_C = = 3 - 1 = 2$

$Q_E = \dfrac{1}{6} \times (1\ 071.8^2 + \cdots + 1\ 143.2^2) = 667\ 962.7$

$S_E = Q_E - P = 2\ 566.4$

$f_E = = 3 - 1 = 2$

$Q_F = \dfrac{1}{6} \times (1\ 074.8^2 + \cdots + 1\ 487.2^2) = 701\ 290.3$

$S_B = Q_B - P = 35\ 893.8$

$f_B = = 3 - 1 = 2$

$Q_N = \dfrac{1}{6} \times (1\ 183.8^2 + \cdots + 1\ 059^2) = 667\ 739.4$

$S_N = Q_N - P = 2\ 342.9$

$f_B = = 3 - 1 = 2$

Calculating S_D is a new problem, and factor D is actually only two levels, level one and level two. The number of these two levels is not the same, the first level of 12 tests, the second level of 6 tests

$K_1^D = 1\ 213.8 + 1\ 120.5 = 23\ 343.3$

$K_2^D = 1\ 126.5, r_1 = 12, r_2 = 6,$

$Q_D = \dfrac{1}{12} \times 2\ 334.3^2 + \dfrac{1}{6} \times 1\ 126.5^2 = 665\ 580.1$

$S_D = Q_D - P = 183.6$

$f_D = 2 - 1 = 1$

From the calculation of S_D we see that the quasi-level has the following characteristics, the degree of freedom is less than that of the orthogonal table. Factor D occupies column 5 of $L_{18}(6^1 \times 3^6)$, but its degree of freedom, $f_5 = 1$, is less than the degree of freedom, $f_8 = 2$, of column 5. In other words, although D accounted for the fifth column, but not filled, not filled the place is the experimental

Chapter 3 Variance Analysis of Orthogonal Design

error, we use S'_e said.

There are two ways to calculate, one is the first five columns to calculate, which is

$Q'_D = \dfrac{1}{6} \times (1\,213.8^2 + 1\,126.5^2 + 1\,120.5^2) = 666\,305.1$

$S'_D = Q'_D - P = 666\,305.5 - 665\,396.5 = 909$

$S'_e = S'_D - S_D = 909 - 183.6 = 725.4$

$f'_e = 1$

the other may is:

$S'_e = S'_T - S_A - S_B - S_C - S_D - S_E - S_F - S_T = 72\,416 - 14\,306.4 - 3\,622.2$
$\quad - 12\,776.2 - 183.6 - 35\,893.8 - 2\,566.4 - 2\,342.9 = 725.4$

Explain the calculation is correct. Two methods depending on the convenience of selection.

An analysis of variance is given in Table 3.2.13. Due to its small size S_D, combine it with the error. From the result of variance analysis, we can get:

a) The influence of cone angle on the cone strength is particularly significant. The influence of the number of ducts is remarkable. No other factors can see any effect.

Table 3.2.13 Variance analysis table

Variation source	Sum of squares	Freedom	Mean square	F value	Critical value
A	$S_A = 14\,306.4$	5	2 861.3	3.52	$F_{0.05}(5,4) = 6.26$
B	$S_B = 3\,622.2$	2	1 811.1	2.23	$F_{0.05}(2,4) = 6.94$
C	$S_C = 12\,776.2$	2	6 388.1	7.86	$F_{0.01}(2,4) = 18.00$
D	$S_D = 35\,893.5$	2	17 946.9	22.07	
E	$S_E = 35\,893.8$	2	1 283.2	1.58	
F	$S_F = 2\,566.4$	1	}813		
Empty column	$S_E = 725.4$	1			
Error	$S_N = 2\,342.9$	2			
SUM	$S_T = 72\,416$	17			

b) Take the angle (E) of the bottom of the bucket to take the E_3 number of air duct (C) as C_1, which is consistent with the visual analysis. Other factors according to the needs of the test within the scope of any one level.

c) The test error in this case is $\sqrt{813} = 28.5\,\mathrm{kg/cm^2}$ accuracy is poor.

Discuss:

a) From the analysis of variance in this case, we see that after the analysis of variance, all the factors are not significant how to do? The reason for this phenomenon is mainly the error is too large to cover up the authenticity of things. Second, if the level of the selected factors there is no difference, it will produce this phenomenon. For the former, one remedy can be taken: Combining some less than or nearly error 5: with it to test for larger S. It is to combine some S with or near the error to merge S_e. In this case, if we use $S_e = S_N + S'_e$ to examine all the factors, they are all insignificant and will be merged into S_e. $S_e = S_n + S'_e + S_D$ The result of the test is that the factor E is particularly significant and the factor C is significant.

Tab 3.2.14 $L_{18}(3^7)$ Test scheme and calculation results

Experimental number	A. Feed grain size	B. Feed grain volume(t/h)	C. Loading capacity(t)	D. gradation $D_{max}:D_{max}$	E. Slurry concentration(t/h)	6	7	Finess modulus of finished sand
	1	2	3	4	5	6	7	
1	2(Mid stone)	3(74.67)	3(23.36)	1(100:0)	1(64.93)	2	2	4.06
2	2(Mid stone)	2(61.45)	3(23.36)	1(100:0)	2(32.55)	1	3	3.87
3	3(Big stone)	1(6.19)	2(27.18)	1(100:0)	3(14.4)	2	3	2.57
4	3(Big stone)	3(74.57)	2(27.18)	1(100:0)	2(36.43)	3	1	4.28
5	1(Small stone)	1(41.31)	1(30.90)	1(100:0)	1(50.39)	1	2	3.42
6	1(Small stone)	2(66.47)	1(30.90)	1(100:0)	3(14.54)	3	1	3.48
7	1(Small stone)	3(73.67)	3(23.54)	3(40:60)	3(16.83)	3	3	3.59
8	1(Small stone)	1(55.07)	3(23.54)	3(40:60)	2(28.63)	2	1	3.91
9	2(Mid stone)	2(66.56)	2(27.14)	3(40:60)	3(14.74)	1	1	3.43
10	2(Mid stone)	1(58.21)	2(27.14)	3(40:60)	1(50.53)	3	2	3.87
11	3(Big stone)	3(75.54)	1(30.68)	3(40:60)	2(36.72)	1	2	3.53
12	3(Big stone)	2(69.43)	1(30.68)	2(40:60)	1(58.72)	2	3	3.91
13	3(Big stone)	2(67.63)	3(22.99)	2(40:60)	1(58.42)	3	1	4.10
14	3(Big stone)	1(58.75)	3(22.99)	2(70:30)	3(14.32)	1	2	2.90
15	1(Small stone)	2(65.58)	2(26.25)	2(70:30)	2(32.63)	2	2	4.41
16	1(Small stone)	3(75.46)	2(26.25)	2(70:30)	1(64.52)	1	3	4.54
17	2(Mid stone)	1(58.33)	1(30.39)	2(70:30)	2(28.32)	3	3	3.29
18	2(Mid stone)	3(75.93)	1(30.39)	2(70:30)	3(16.40)	2	1	2.57
K_1	23.35	19.96	20.2	21.68	23.90	21.69	21.81	
K_2	21.09	23.20	23.10	21.81	23.29	21.43	22.25	SUM:65.73
K_3	21.29	22.57	22.43	22.24	18.54	22.61	21.77	
R	2.26	3.24	2.90	0.56	5.36	1.18	0.54	
S_i	0.521 8	0.982 8	0.768 5	0.028 68	2.87	0.128	0.029 4	

Chapter 3 Variance Analysis of Orthogonal Design

b) The above conclusion is different from the original data. We think, the $L_{18}(6^1 \times 3^6)$ mixed level meter must be properly applied to the visual analysis and, if necessary, be further analyzed for variance, even if the error is high, the main problem of the problem can still be captured.

3.2.2.5 Analysis of the Variance of the Orthogonal Table with the Degree of Freedom Unsaturated

The so-called orthogonal table with the degree of freedom unsaturated means that the sum of the degrees of freedom in each column of such tables is less than the total degree of freedom. Such as $L_{12}(3^1 \times 2^4)$, $L_{18}(3^7)$ and so on. A special note to ANOVA for such tables is that you can not directly calculate trial errors using empty columns, even if there is no interaction between the factors.

【Example 3.2.6】 A project using rod mill production into the sand, the impact of artificial sand process conditions on the modulus of fineness.

The test protocol and visual analysis results are shown in Table 3.2.14. The calculation results for each column are listed below the table.

The total square and $S_T = 5.542$.

According to the normal law of the orthogonal table should be
$$S_T = S_A + S_B + S_C + S_D + S_E + S_6 + S_7$$
but here,
$$\begin{aligned} S_T &= S_A + S_B + S_C + S_D + S_E + S_6 + S_7 \\ &= 0.521\,8 + 0.983\,8 + 0.768\,5 + 0.028\,26 + 2.87 + 0.128\,1 + 0.029\,2 \\ &= 5.33 < S_T = 5.542 \end{aligned}$$

This is because the orthogonal table $L_{18}(3^7)$ degree of freedom is not saturated. The table has a total degree of freedom $f_T = 18 - 1 = 17$, and the degree of freedom in each column is the number of levels minus one, that is $f_A = f_B = f_C = f_D = f_E = f_6 = f_7 = 3 - 1 = 2$, and the sum is 14. $S_{e1} = S_6 + S_7 = 0.157\,3$ for empty columns 6,7 reflect only part of the error, and the other part $S_{e2} = 5.542 - 5.33 = 0.212$, and its degree of freedom $f_{e2} = 17 - 14 = 3$.

An analysis of variance is shown in Table 3.2.15. Due to small S_D knowledge, it is combined with S_{e1}, S_{e2} to jointly estimate the error.

Table 3.2.15 Variance analysis table

Variation source	Sum of squares	Freedom	Mean square	F value	Critical value
A	$S_A = 0.521\,8$	2	0.260 9	5.9	$F_{0.05}(2,9) = 4.63$
B	$S_B = 0.983\,8$	2	0.491 9	11.13	$F_{0.01}(2,9) = 8.00$
C	$S_C = 0.768\,5$	2	0.384 2	8.69	
E	$S_E = 2.87$	2	1.435 0	32.47	
Error	$S_e = 0.398\,1$	9	0.044 2		
SUM	$S_T = 5.542$	17			

Analysis of variance showed that the thick pulp, feed material, and installed on the thin rod. The effect of degree modulus is particularly pronounced; the effect of feed grain size is significant and the rod grade does not influential. The experimental error of this example $\hat{\sigma}_e = \sqrt{\bar{S}_e} = 0.21$.

It can be seen from this example that if we do not know $L_{18}(3^7)$ is an orthogonal table with a degree of freedom and unsaturation, ANOVA will draw the wrong conclusion.

3.3 Error Handling

Some common ANOVA examples listed in 3.2, we see that the difference is a measure of whether the test is significant or not, and its size directly affects the analysis of test results. Therefore, it is necessary to discuss the error.

3.3.1 The Type of Error

In orthogonal design, the error can be divided into experimental error and repeated sampling error. (Or group test error). In the test error can be Fen for the general test error (that is, the error of the orthogonal table empty column), repeated trial and error, the proposed level of error and out-of-column error and so on. For example, only empty columns in Example 3.2.2 provide experimental errors; Example 3.2.3 contains not only null columns but also repeated tests to provide experimental errors; Example 3.2.5 In addition to null columns, there is a quasi-horizontal error test Error; Example 3.2.6 In addition to the null column, there is also an error in the orthogonal out-of-table error to provide the test error w These errors are the overall error.

They can be used alone or in combination to examine the significance of the factors.

Repeated sampling error is a local error. Because repeated sampling is easier than repeated testing, under certain conditions, the significance of the factors can also be checked alone or in combination with experimental errors (Example 3.2.4).

Orthogonal design, we only focus on how to reduce the number of trials and scientific analysis of the test results, there is no discussion of how to reduce the test error. Due to the size of the test error, a direct impact on the analysis of the test results, if the test control conditions are not good, will result in large experimental error. The greater the experimental error, the authenticity of the test can easily be masked, ranging from the need to increase the number of tests, while the result is wrong conclusions. Therefore, in the concrete test, to do; t eliminate all the interference can be ruled out to make the test error minimized. For example, a test may be conducted simultaneously with several instruments or several testers, etc., and different instruments and testers will give errors to the test. The solution is to consider several instruments or several testers as a factor in the column number of the orthogonal table.

3.3.2 The Standard of Merger Error and Impact Rate

In the F value table, when the degree of freedom error is small $(f_e = 1-5)$, fe changes only 1, the F value changes greatly; the error degree of freedom 9 increases gradually $(f_e > 6)$; when the change is 1, the F value changes gradually Smaller Therefore, when the error degree of freedom is small, the number of tests should be taken more often and the case where the degree of freedom is increased should be judged as significant, at which time it is judged as not significant. In the trial design using orthogonal funeral, the degree of freedom of error (null columns) is generally small and, where possible, at least 6 is better. At the same time, when $f_e > 20$, each change in the ruler 1, F value only slightly changes, therefore, 6 in the most ideal between 20. In order to meet the above requirements, in order to improve the accuracy of test results, the errors are often merged.

An analysis of variance for the non-calculated F in Table 3.2.13 of Example 3.2.5 is shown in Table 3.3.1.

Table 3.3.1 Variance analysis table

Variation source	Sum of squares	Freedom	Mean square
A	$S_A = 14\ 306.4$	5	2 861.3
B	$S_B = 3\ 622.2$	2	1 811.1
C	$S_C = 12\ 776.2$	2	6 388.1
D	$S_D = 186.6$	1	183.6
E	$S_E = 35\ 893.8$	2	17 946.9
F	$S_F = 2\ 566.4$	2	1 283.2
Error	$S_e = 3\ 068.3$	3	1 022.8

The criterion of merging is to compare the mean square of the main effect in Table 3.3.1 with the mean square of the error term (1 022.8 in this case) and will be considered as the same magnitude (roughly 2 – 3 times or so) The main effect can be regarded as error, re-analysis of variance in Table 3.3.2.

Table 3.3.2

Variation source	Sum of squares	Freedom	Mean square	F value	Critical value
A	$S_A = 14\ 306.4$	5	2 861.3	2.42	$F_{0.01}(2,8) = 8.7$
C	$S_C = 12\ 776.2$	2	6 388.1	5.4	$F_{0.05}(2,8) = 4.5$
E	$S_E = 35\ 893.8$	2	17 946.9	15.21	$F_{0.10}(2,8) = 2.7$
Error	$S_e = S_n + S_e' + S_B + S_D + S_F$ $= 9\ 440.5$	8	1 180.1		

The error in Table 3.3.2 is called the error after merging. Therefore, the factor becomes the main effect A, C, E and the error after the merger. So come to judge: the factor E is particularly significant, because significant. From this it can be seen that there are times when the test is carried out after the merger and sometimes the test without the merger. In this case, the factors need to be divided into two situations:
1. factors that are always significant (such as factor E);
2. Due to different test methods, sometimes become significant, and sometimes become insignificant (for example, factor C). In the first case, the optimal process level must be chosen on the basis of the principle of preference, while in the second case, the level of expertise or influence should be

used. In the meantime, it is best to study repetitive tests, the number of levels and their amplitude, etc., when arranging further tests. The following discussion focuses on impact.

The so-called shadow rate is the impact of various factors on the test results the proportion of the number. Factors affect the rate of calculation according to the following formula:

$$\rho_i(\%) = \frac{S_i - \frac{S_e}{f_e}f_i}{S_T} \times 100 = \frac{S_i - \overline{S_e}f_i}{S_T} \times 100$$

which in:

S_i and f_i = represent the sum of squares and degrees of freedom of the factors, respectively

S_e = Mean square error

S_t = sum of squares.

Adulteration of various techniques of fly ash Feng Feng in the Table 3.3.3. Error color response $\rho_e = 100 - \rho_A - \rho_B - \rho_V - \rho_B - \rho_P = 19.1\%$

Table 3.3.3

Factor	Sum of squares S	Freedom f_i	Influence ratio $\rho(\%)$
A	$S_A = 14\ 306.4$	5	14.1
B	$S_B = 3\ 622.2$	2	2.8
C	$S_C = 12\ 776.2$	2	15.4
E	$S_E = 35\ 893.8$	2	47.3
F	$S_F = 2\ 566.4$	2	1.3
Error	$S_e = 3\ 251.9$	4	19.1
S_T	$S_T = 7\ 2416$	17	

The calculation results show that the factor E accounts for 47% of the total variation, followed by C and A, each accounting for about 15%, the second is B and F, and the influence degree is below 10%.

Once you know your impact, you can interpret the results and decide whether or not you need to take action on them.

3.3.3 Test Level

In concrete testing, the advantages and disadvantages of setting a test result g, test error is an important indicator. Earlier pointed out that the test of poor estimates, with the mean square error $\hat{\sigma} = \sqrt{\overline{S_e}}$; said the different assessment indicators mean value range within the change. Practice shows that different assessment indicators, the coefficient of variation (Macro for the assessment of the insult the average) remained the same number, under normal $C_v < 5\%$ or less. Based on current experience, it is preliminarily believed that the According to the existing experience, it is preliminarily believed that according to the size of the Cv value, the test level can be divided into three, that is, $C_v < 5\%$ belongs to the superior class; $C_v = 5 - 10\%$ belongs to the general level; $C_v > 10\%$ belongs to the bad. For example, $x = 1\ 639/9 = 181.8$, $\hat{\sigma}_e = 7.2$, $C_v = 4.0\%$; Example 3.2.4, $C_v = 5.2\%$; example 3.2.5, $C_v = 14.8\%$; Example 3.2.6, $C_v = 5.7\%$

It can be seen that the test levels of Example 3.2.2 and Example 3.2.3 are superior; Example 3.2.4 and Example 3.2.6 are general; and Example 3.2.5 is poor. These data show that, in spite of the large error fluctuations in concrete tests, only the test conditions are controlled, and in most cases the test level can be well above the water level. Example 3.2.5 larger test errors may be related to the conditions of field test.

Chapter 4
Simple Variance Analysis of Orthogonal Design and Multiple Comparisons

ANOVA calculations do not make too much effort But since the box calculates the sum of squares of a series of data, it is "still troublesome" in the absence of computational tools. Naturally, it is desirable to find an easy way to use the difference. The purpose of this method is to calculate relatively simply and its efficiency is similar to that of usual ANOVA. It is worth recommending to practical workers. In addition, it is often translated into practical work. When one, one, is overwhelmed by the presence of a jerk, one can judge whether this is significant because of jerkiness. Which is not significant, in order to seek more reasonable process conditions, the so-called multiple comparisons. This chapter will be combined with examples to introduce how to use range method for analysis of variance and multiple comparisons.

4.1 Extreme and Mean Squares

In pant 3.1, we introduced the formula for calculating sample mean $\overline{X}$ and mean square error S:

$$\overline{x} = \hat{\mu} = \frac{1}{n} \sum_{i=1}^{n} x_i$$

$$S = \hat{\sigma} = \sqrt{\frac{1}{n-1} \sum_{i=1}^{n} (x_i - \overline{x})^2}$$

They are estimated as the overall mean u and the overall mean square error. However, since the calculation of S saw the square σ of a series of data, the calculation is large, and the calculation can be simplified with the very poor. The formula for estimating the overall mean square error with very poor σ is:

$$\hat{\sigma} = \frac{1}{d_2} \qquad (4.1.1)$$

The variation of d_2 varies with the sample size, n, and is listed in the general mathematical statistics (value for search. The conditions for using this estimate are:
1. n can not be too big. When $n > 12$, the accuracy is poor. If large

Chapter 4 Simple Variance Analysis of Orthogonal Design and Multiple Comparisons

When the data can be divided into several groups, so that each group of data is less than 12, then use an average of $\bar{R}$ instead of R.

2. The number of groups can not be too small. When the estimation accuracy is higher, it is best to have 20 groups.

In the analysis of variance, the first condition (n is equal to the number of levels is generally satisfied, but the second is generally not satisfied. In order to estimate the grain size,

The d_2 table is modified into $d(n, l)$ table (see Appendix III Table IV). To illustrate the table use method, look at two simple examples.

【Example 4.1.1】 From the normal population taken three compressive strength samples: 215, 230, 251 kg/cm², trying to estimate the mean square error.

According to Equation 3.1.3 and 3.1.6:

$$\bar{x} = \frac{1}{3} \times (215 + 230 + 251) = 232$$

$$S = \hat{\sigma} = \sqrt{\frac{(215-232)^2 + (230-232)^2 \times (251-232)^2}{3-1}} = 18.1$$

With very poor estimation, the three samples are very poor:

$$R = 251 - 215 = 36$$

Check $d(n, l)$ table, the table has two parameters, n represents the number of samples within each group, l said the number of groups. Here $n = 3, l = 1$, the corresponding table 1.91, with $d(n, l)$ instead of d_2, into the Equation 4.1.1:

$$\hat{\sigma} = R/d(3,1) = 36/1.91 = 18.8$$

Similar to 18.1. If you use d_2, in the last column of $d(n, l)$ table

$L = \infty$, It is d_2. When $n = 3$, $d_2 = 1.693$, then there

$$\hat{\sigma} = R/d_2 = 36/1.693 = 21.2$$

It can he seen that the difference between $\hat{\sigma}'$ and S is longer than that of $\hat{\sigma}$ and S. when the number of L is small. the effece of replacing d_2 with $d(n,l)$ is good

【Example 4.1.2】 Example 3.1.2 of the range method is used to calculate the mean square error of ten compressive strengths 214, 227, 217, 231, 224, 241, 223, 249, 236.

In front of $S = 14.07 \text{kg/cm}^2$. The difference of ten data $R = 258 - 214 = 44$, in this case, $l = 1$ n $= 10, d(n,l) = d(10,1) = 3.18$, then

$$\hat{\sigma} = R/d(10,1) = 44/3.18 = 13.84$$

It is close to $S = 14.07$

For the general situation, if divided into l group, there are n samples in each group, and the difference in group l is $R_1, R_2, \cdots, R_l$, respectively, and its average difference

$$\bar{R} = \frac{1}{l} \sum_{i=1}^{l} R_i \qquad (4.1.2)$$

The mean square error is estimated:

$$\hat{\sigma} = R/d(n,l) \qquad (4.1.3)$$

When l is sufficiently large, $d(n, l) = d_2$. In addition to $d(n,l)$ in Table III of Appendix III, a table of numerical values is also presented that shows the degree of freedom for statistical tests. It is usually rounded off in use. There is an approximation of the degree of freedom:

$$\varphi(n,l) \approx 0.90l(n-1) \tag{4.1.4}$$

4.2 Application of Range Method in Variance Analysis

4.2.1 The Calculation Method

The following Example 3.1.1 data, simplify the analysis of variance.

1. Calculate the average intensity $\overline{X}$ at each dosage and the range R, listed in Table 4.2.1.

For example, the odds ratios at P_1 are 45, 24, 52, 60 respectively, that is, $R_1 = 60 - 24 = 36$, and so on.

Table 4.2.1

Repeated test times	Strength at all levels $x_i = R_{28} - 140$			
	P_1	P_2	P_3	P_4
1	45	33	14	5
2	24	8	0	0
3	52	40	29	18
4	60	34	30	28
$\sum x_i$	181	115	73	51
Average X_i	45.25	28.75	18.25	12.75
Range R_i	36	32	30	28

2. Estimate the test error

Find the average value of four very poor R

$$\overline{R} = \frac{1}{4} \times (R_1 + R_2 + R_3 + R_4)$$
$$= \frac{1}{4} \times (36 + 32 + 30 + 28)$$
$$= 31.5$$

There are four groups, four groups of data, that is $l = 4, n = 4$, look-up table $d(4,4) = 2.11$, by the Equation 4.1.3, the error of the mean square error is calculated as

$$\hat{\sigma} = R/d(4,4) = 31.5/2.11 = 14.9$$

The square of $\hat{\sigma}$ is the approximate value of $\overline{S}_e$, $\hat{\sigma}^2 = 20$ and $\overline{S}_e = 200.4$ (Seen in Table 3.1.4) is obtained by square sum method. The degree of freedom of error term is $\varphi(4,4)$, $f_e = \varphi(4,4) = 11.2 \approx 11$, And when square sum is used to calculate $f_e = 12$, It shaws that the satistility of the method is sligitly reduced.

3. Test mixed although the impact of the intensity is significant

Test in two ways:

a. *F* test

Similarly, S_A is also estimated in a very poor way. The difference a R_A of the four averages of the four quantities is calculated and r_A is the range of 45.25, 28.75, 18.25, 12.75, ie $r_A = 45.25 - 12.75 = 32.5$. There is only one group, so $l = 1$, the group has four numbers, that is, check $d(4,1) = 2.24$. So the approximation of $\bar{S}_A$

$$\bar{S}'_A = r[r_A/d(4,1)] \tag{4.2.1}$$

Where r is the number of trials m for each level, where $r = 4$. Substituting the data into Equation (4.2.1) yields $S'_A = 4 \times [32.5/2.24]^2 = 842$, thereby obtaining

$$F_A = \frac{\bar{S}'_A}{R} = \frac{842}{222} = 3.8$$

Therefore, the effect of dosage on strength is significant. This is in good agreement with the conclusions in Table 3.1.4. This shows that, with extremely poor method, skip the sum of square sum, it is appropriate to estimate the mean square, much more convenient to calculate $f_A = \varphi(4,1) = 2.9 \approx 3$, From table, $F_{0.01}(3,11) = 6.2$, $F_{0.05}(3,11) = 3.6$.

b. Comparison of the width of the *T* method of calculation

$$q_A = \sqrt{r} \cdot r_A/\hat{\sigma} \tag{4.2.2}$$

Compare it with the value of $q_a(P, h)$ on the third column of Appendix III, where P is the number of levels and f_e is the degree of freedom of error. This case $P = 4$, $f_e = 11$, look-up table

$$q_A = \sqrt{4} \cdot 32.5/14.9 = 4.36$$

$q_A > q_{0.05}$, The effect of the amount of admixture is significant, consistent with the *F* test. Finally, we need to explain the two points.

1. In order to reduce the amount of calculation, you can not calculate the average of each level, only calculate the sum, remember $K_1, K_2, \cdots, K_p$ the difference is R_A, then

$$r_a = \frac{1}{r}R_A \tag{4.2.3}$$

So Formulas 4.2.1 and 4.2.2 become:

$$\bar{S}'_A = \frac{1}{r}[R_A/d(p,1)]^2 \tag{4.2.4}$$

$$q_A = \frac{1}{r}R_A/\hat{\sigma}_e \tag{4.2.5}$$

2. In terms of very poor, the two test methods to *T* method are better as low. Decreased sensitivity test more. The *T* method can also be flat on the level multiple comparisons of means, which will be detailed later.

4.2.2 Calculate the Example

【Example 4.2.1】 In Example 3.2.1, a routine ANOVA was performed on Example 1.2.1 (trial-made strong concrete), which can be solved more easily with extremely poor itchiness. Transcribe Table 3.2.1 and list it in Table 4.2.2.

Table 4.2.2

Experimental number	A	B	C		$X_i = R_{28} - 830$
	1	2	3	4	
1	1 (10%)	1 (2%)	1 (3%)	1	-65
2	1	2 (3.5%)	2 (6%)	2	-20
3	1	3 (5.0%)	3 (9%)	3	-72
4	2 (15%)	1	2	3	27
5	2	2	3	1	61
6	2	3	1	2	-65
7	3 (20%)	1	3	2	77
8	3	2	1	3	37
9	3	3	2	1	30
K_1	-157	39	-93	26	
K_2	23	78	37	-8	SUM = 10
K_3	144	-107	66	-8	
R	301	185	159	34	

1. Estimated test error Let ρ represent the number of columns of the orthogonal error table, n is the number of horizontal columns, r is the number of tests with these columns at the same level, the test error of the mean square error cr is estimated

$$\hat{\sigma} = \overline{R_i} / d(n,l) / \sqrt{r} \tag{4.2.6}$$

which in: In the formula, $l = 1, n = 3, r = 3$

$\overline{R} = R_4 = 34$

$d(n,l) = d(3,1) = 1.91$

Substitute data into Equation (4.2.6):

$\hat{\sigma} = 34/1.91/\sqrt{3} = 10.3$

$\sigma^2 = 10.3^2 = 105.9$

and Table 3.2.2 are similar as $\overline{S}_e = 128.5$. Each degree of freedom

$f_e = \varphi(n,l) = \varphi(3,1) = 2$

2. Test A, B, C significant

Let R_A denote the very poor result of the column RA for each level of test. The test for each level: the number of times, P the level of A.

$R = KR^m$

$$q_A = \frac{R_A}{\sigma \sqrt{r_A}}$$

Use it with the q table on the $q_a(p, f_e)$ to determine the significance of comparison. Table 4.2.2, $R_A = 301$, then

$q_A = 301/10.3/\sqrt{3} = 16.9$

Check the heart by the q table. $q_{0.05}(3,2) = 8.337$ and $q_{0.01}(3,2) = 19.0$ because of $q_{0.01} > q_A > q_{0.05}$ so factor A has a significant impact on the intensity A.

Use the same method to test B, C.

$q_B = 185/10.3/\sqrt{3} = 10.4$,

$q_C = 159/10.3/\sqrt{3} = 8.9$,

Can be seen, the impact of factors B and C on the strength is also significant.

For the word level, using F to test A, B, C of the significance of relatively simple, the approximate value of $\bar{S}_A$

$$\bar{S}_A' = [R_A/d(p,1)/\sqrt{r_A}]^2$$

In this case $d(p,1) = d(3,1) = d(n,l)$, $r_A = r = 3$

or

$$F_A = \frac{\bar{S}_A'^2}{\hat{\sigma}^2}$$

Can be simplified as

$$F_A = \frac{R_A^2}{R_i^2}$$

Brought into the data:

$$F_A = \frac{301^2}{342^2} = 78.37$$

$$F_B = \frac{185^2}{34^2} = 29.6$$

$$F_e = \frac{159^2}{34^2} = 21.58$$

The degree of freedom due to the cable $f_A = f_B = f_C = \varphi(3,1) = 2$, check! Table F, then $F_{0.01}(2,2) = 99.0$ and $F_{0.05}(2,2) = 19.0$, so the significant conclusion is basically the same as Table 3.2.2, and the calculation is simpler.

【Example 4.2.2】 For Example 3.2.2 (DH water reducer test) to simplify the calculation of poor results see Table 3.2.4.

1. Estimate the test error

In this case, the orthogonal table has two empty columns, that is, $l = 2, n = 3, r = 3$ Then,

$\bar{R}_i = 0.5(R_3 + R_4) = 22.5$

$d(n,l) = d(3,2) = 1.81$

$f_e = \varphi(3,2) = 3.8 = 4$

Then $\hat{\sigma} = \bar{R}_i/d(3,2)/\sqrt{r} = 22.5/1.81/\sqrt{3} = 7.19$ and Burgundy in Table 3.2.5, $S_e = 52.5$ similar.

2. Test the significance of A, B.

$R_A = 55$ From Formulas 4.2.7 we can get:

$F_A = 55^2/22.5^2 = 5.98$

$R_b = 81$, then

$F_B = 81^2/22.5^2 = 12.96$

Their degree of freedom $f_A = f_B = \varphi(3,1) = 2$ Check from table F.

$F_{0.01}(2,4) = 18.0$, $F_{0.05}(2,4) = 6.9$, $F_{0.1}(2,4) = 4.3$

Orthogonal Design in Concrete Application

The significance of the conclusion is completely different from Table 3.2.5.

【Example 4.2.3】 In order to compare the difference between the moisture content of stone surface by drying and drying method, six skilled testers were selected for the surface water content of medium stone and small stone respectively by the above two methods For the measurements, we use the data obtained from the experiments to make the simplified variances for orthogonal designs.

The results of the test and the calculation of the level and range are shown in Table 4.2.3 and Table 4.2.4.

Table 4.2.3 Factor level table

Factor	Level					
	1	2	3	4	5	6
A. Testers	1#	2#	3#	4#	5#	6#
B. Drying method	roast	blow				
C. Stone size	Middle stone	Small stone				

1. Estimate the test error

The difference between the five water-string repeats of the same test reflects the experimental error and the last column of Table 4.2.4 reflects this error. Average them:

Table 4.2.4 $L_{12}(6 \times 2^2)$ Test results and calculation

Experimental number	A	B	C	5 test results of water conent(%)					SUM	R_i
	1	2	3	1	2	3	4	5		
1	1(1#)	1(roast)	1(middle)	0.19	-0.01	0.16	0.24	0.36	0.94	0.37
2	2(2#)	1	2(small)	0.53	0.71	0.34	0.37	0.47	2.42	0.37
3	1	2(blow)	2	0.52	0.54	0.60	0.59	0.47	2.72	0.13
4	2	2	1	0.27	0.22	0.17	0.20	0.20	1.06	0.10
5	3(3#)	1	2	0.73	0.73	0.48	0.65	0.65	3.24	0.25
6	4(4#)	1	1	0.80	0.86	0.43	0.31	0.53	1.98	0.23
7	3	2	1	0.21	0.23	0.25	0.18	0.4	1.21	0.16
8	4	2	2	0.77	0.54	0.55	0.57	0.67	3.1	0.23
9	5(5#)	1	1	0.17	0.21	0.22	0.42	0.44	1.6	0.27
10	6(6#)	1	2	0.73	0.73	0.84	0.61	0.31	2.72	0.42
11	5	2	2	0.59	0.68	0.59	0.76	0.42	2.99	0.34
12	6	2	1	0.26	0.19	0.14	0.27	0.18	1.04	0.13
K_1	3.66	12.71	7.64						Σ = 24.83	Σ = 3.00
K_2	3.48	12.12	17.19							
K_3	4.45									
K_4	5.03									
K_5	4.45									
K_6	3.76									
R	1.55	0.59	9.55							

$$\bar{R} = \frac{1}{12} \times (0.37 + \cdots + 0.13) = 0.25$$

Here $l = 12, n = 5$ and check $d(5,12) = 2.339$ then

$\hat{\sigma} = \overline{R}/d(5,12) = 0.25/2.339 = 0.1069$

from the Formula 4-1-4 then:

$\varphi(n,l) = 0.9l(n-1) = 44$

Due to a missing data, the error degree of freedom should be reduced by 1.

$f_e = 44 - 1 = 43$

2. Test A, B, C significance

To factor A,

$R_A = 1.55, r_A = 2 \times 5 = 10$

$q_A = R_A/\hat{\sigma}/\sqrt{r_A} = 1.55/0.1089/\sqrt{10} = 4.586$

to factor B and C,

$R_B = 1.55, R_C = 9.55, r_B = r_C = 6 \times 5 = 30$

$q_B = 0.59/0.1069/\sqrt{30} = 1.008$

$q_C = 9.55/0.1069/\sqrt{30} = 16.313$

$q_A > q_{0.05}(6,43) = 4.23$ A significance

$q_B > q_{0.1}(2,43) = 2.38$ can not see B has influence

$q_C > q_{0.10}(2,43) = 3.82$ C particular significance

From the result of variance analysis, we get:

1. Main Inference on the moisture content of the primary and secondary annoying sequence for the stone size →test person→dry methods.

2. Stone size has a particularly significant impact on moisture content In the stone size and test methods under certain conditions, the tester's operating level significantly affected the moisture content of the "two test methods for measuring the moisture content results Can not see any impact.

Discussion: In this case, the original author did not arrange the experimental design according to orthogonal design and made a total of 120 glycosides. According to the requirements of the orthogonal table, we choose 60 trial results to simplify the analysis of variance. The conclusion is different from the original author. If we use common variance analysis to process the data in this case, the calculation of working love is much larger.

From these examples we further see that in the Orthogonal Design for M-seized. When testing, using the method of variance than the commonly used ANOVA to be simple and much more convenient. Things are always divided into two parts. The difference method has advantages and drawbacks. For example, it can not directly estimate the total sum of squares, which is slightly less efficient than the sum of squares method. And if the model is different, when the factors have an interaction, can not use $d(n.l)$ table, these are its shortcomings. Readers can accord their own situation to be adopted.

4.3 Multiple Comparison of T Method

4.3.1 The Meaning and Methods

In the previous section, we introduced the T-test using the F test and the multiple S comparisons to determine whether the influence of the citation is significant. When the factors are significant, it does not mean that all the differences between the levels are significant. What is the difference between the levels of pinch significant or insignificant, for practical problems is of great significance. As mentioned earlier, in the orthogonal design, the best choice of the level of K is the highest value of all, such as the compressive strength is as high as possible. However, the process conditions or mixes selected by this principle are sometimes difficult to realize in production due to the objective conditions, and sometimes the cost is too high. Naturally at this time it would be desirable to replace it with other levels of insignificance that are not significantly different from the optimal level, which are easy to produce or less costly. As shown in Example 3.1.1 variance analysis results show that the amount of super plasticizer has a significant impact on compressive strength. Now to further test the four kinds of holding; t (four levels) compressive strength, which levels have significant differences? What is not significant difference between the water? A type of problem is called multiple comparisons. There are several ways to solve multiple comparisons. This book only describes the use of the q-table as a test: T. Readers who want to know more about this method can consult [1].

The basic method of multiple comparisons is described below in conjunction with Example 3.1.1. For multiple comparisons, it is necessary to know the following parameters:

m = represents the number of levels to be compared

r = represents the same number of experimental repetitions

$\overline{S}_e$ = represents the square of expeimeatal error

$f_e = \varphi(n,1)$: represents the square of experimental error, represents m levels of average compressive strength;

In this case,

$m = 4, n = 4, \overline{S}_e = 200.4, f_e = 12$

$\overline{K}_1 = 185.25, \overline{K}_2 = 168.75, \overline{K}_3 = 158.25, \overline{K}_4 = 152.75$

T method to compare, we should use to judge. q table has two parameters m and the number of levels to be compared, φ is the degree of freedom, for example $q_{0.05}(m,\varphi) = q_{0.05}(4,12) = 4.2$. Compare any two kinds of doping the amount of (level > average compressive strength elbow, to determine the scale of

$$d_r = q_{0.05}(m,\varphi)\sqrt{\frac{\overline{S}_e}{r}} \quad (4.3.1)$$

If the average compressive strength difference is less than d_T. There is no significant difference between the two admixtures (horizontal); if the difference between the average compressive strength is greater than dr, there is a significant difference between the two levels (level).

$$d_r = q_{0.05}(4,12)\sqrt{\frac{\overline{S}_e}{r}} = 4.2\sqrt{\frac{200.4}{4}} = 29.7$$

In this case, the difference between the average compressive strength of any two kinds of content is:

$$d_{12} = |\overline{K}_1 - \overline{K}_2| = |185.25 - 168.75| = 16.5$$

$d_{13} = |\overline{K}_1 - \overline{K}_3| = |185.25 - 158.75| = 26.5$

$d_{14} = |\overline{K}_1 - \overline{K}_4| = |185.25 - 152.75| = 32.5$

$d_{23} = |\overline{K}_2 - \overline{K}_3| = |168.75 - 158.25| = 10.5$

$d_{24} = |\overline{K}_2 - \overline{K}_4| = |168.75 - 152.25| = 16.0$

$d_{34} = |\overline{K}_3 - \overline{K}_4| = |158.25 - 152.75| = 5.5$

Visible, only one of them $K_1 K_2 K_3 K_4$, $d_{14} = 32.5 > d_T = 29.7$ Others are less than d_T. Therefore, only the amount? P_1 and P_4 between the average compressive strength different, the rest were no significant differences. This gives us plenty of room for choosing the optimal level.

4.3.2 Examples

【Example 4.3.1】 Test Example 3.2.2 three naphthalene super plasticizer products are significant differences?

In this case
$$\overline{S}_e = 52.5, f_e = 4$$

Factor B (water reducer species), $m = 3, r = 3$

$\overline{K}_1 = 168, \overline{K}_2 = 183, \overline{K}_3 = 195$,

Check Q can get $q_{0.05}(m, \varphi) = q_{0.05}(3,4) = 5.04$, based on Formula (4.3.1),

$$d_T = q_{0.05}(3,4)\sqrt{\frac{\overline{S}_e}{r}} = 5.04\sqrt{\frac{52.5}{3}} = 21.0$$

Calculate:

$d_{12} = |\overline{K}_1 - \overline{K}_2| = |168 - 183| = 15$

$d_{13} = |\overline{K}_1 - \overline{K}_3| = |168 - 195| = 27$

$d_{23} = |\overline{K}_2 - \overline{K}_3| = |183 - 195| = 12$

Can be seen, naphthalene super plasticizer DH-3 only with DH-1 significant difference, while DH-3 and DH-2 no significant difference. From the existing test data analysis: DH-sub-better than DH-1, and it is not more than the performance of DH-2 showed no significant difference. Therefore, the average strength of DH-3-containing concrete obtained in Example 3.2.2 is only a superficial phenomenon compared with that of DH-2-doped concrete.

【Example 4.3.2】 Test Example 3.2.4 in the vibrating conditions between the various levels of Kun identity?

In this case, $\overline{S}_e = 534.616, f_e = 18$

For factor A (vibration condition), $m = 4, r = 6, \overline{K}_1 = 363, \overline{K}_2 = 446, \overline{K}_3 = 472, \overline{K}_4 = 480$

Check q can get $q_{0.05}(4,18) = 4.0$ based on Formula (4.3.1),

$$d_T = q_{0.05}(4,18)\sqrt{\frac{\overline{S}_e}{r}} = 37.8$$

Calculate:

$d_{12} = |\overline{K}_1 - \overline{K}_2| = |363 - 446| = 83$

$$d_{13} = |\overline{K}_1 - \overline{K}_3| = |363 - 472| = 109$$
$$d_{14} = |\overline{K}_1 - \overline{K}_4| = |363 - 480| = 117$$
$$d_{23} = |\overline{K}_2 - \overline{K}_3| = |446 - 472| = 26$$
$$d_{24} = |\overline{K}_2 - \overline{K}_4| = |446 - 480| = 35$$
$$d_{34} = |\overline{K}_3 - \overline{K}_4| = |472 - 480| = 8$$

Only wide, with $\overline{K}_1, \overline{K}_2, \overline{K}_3, \overline{K}_4$ significant differences between the other levels were not significantly different. This means that there is a significant difference between the average intensity of the vibrations and the intensity of the vibrations or the plugs, while the intensities obtained for the vibrations of 15 s, 30 s and interpolating are not significantly different. That tamping can effectively improve the strength of the flow of concrete, it must be tamping. As for the vibration 15 s or 30 s or plug-in, according to the specific circumstances.

【Example 4.3.3】 The level of operation between testers in Comparative Example 4.2.3 $S_e = \hat{\sigma}^2 = 0.0114, f_e = 43$

There is no significant difference. For factor A, $m = 6$, $r = 2 \times 5 = 10$
$\overline{K}_1 = 0.366, \overline{K}_2 = 0.348, \overline{K}_3 = 0.445, \overline{K}_4 = 0.503, \overline{K}_5 = 0.445, \overline{K}_6 = 0.376$
based on formula 4.3.1,

$$d_T = q_{0.05}(6, 43) \sqrt{\frac{S_e}{r}} = 0.143$$

Calculate:
$$d_{12} = |0.366 - 0.348| = 0.018$$
$$d_{13} = |0.366 - 0.445| = 0.079$$
$$d_{14} = |0.366 - 0.503| = 0.137$$
$$d_{15} = |0.366 - 0.445| = 0.079$$
$$d_{16} = |0.366 - 0.376| = 0.01$$
$$d_{23} = |0.348 - 0.445| = 0.097$$
$$d_{24} = |0.348 - 0.503| = 0.155$$
$$d_{25} = |0.348 - 0.445| = 0.107$$
$$d_{26} = |0.348 - 0.376| = 0.028$$
$$d_{34} = |0.445 - 0.503| = 0.058$$
$$d_{35} = |0.445 - 0.445| = 0$$
$$d_{36} = |0.445 - 0.376| = 0.069$$
$$d_{45} = |0.503 - 0.376| = 0.127$$
$$d_{56} = |0.445 - 0.376| = 0.069$$

Only $d_{24} > d_T$, others are less than d_T. That is, there is a significant difference between the moisture content measured by 2# and 4# testers, while there is no significant difference between the others.

Multiple comparisons show that in order to obtain the results of the water content of small and medium-sized stones, the level of operation of the testing personnel should be wider than that of the testing personnel.

We have only selected a few individual cases of the above significant factors, respectively, in the level of multiple comparisons, designed to illustrate the multiple comparison method in solving practical problems in the practical value. In fact, when choosing the optimal process conditions or optimal mix ratios, we can make multiple m comparisons between levels of any significant factor, based on the required fit. From here we can see that it is hard to tell. Analysis of variance based on the analysis, and then makes multiple comparisons, making the problem analysis more profound. The author believes that the multiple comparison method applied to the analysis of concrete test data can significantly improve the quality of the test results in order to be more economical and reasonable to choose a good process or matching conditions.

Chapter 5
The Orthogonal Design of Interaction Function

If the purpose of the test is not only to seek better formulations or processes, but also to get a more accurate understanding of the impact of various factors on the size of the indicator, especially the different levels of various factors with each other on the index, you need to consider the factors of the interaction, up to the concrete materials may be encountered in the experimental research-class issues. It is difficult to determine the magnitude of the interaction when using the "isolated variable s method" for comparison, and the orthogonal design can consider the interaction and gives a size estimate, which is one of its main advantages. This chapter will introduce the concepts and judgments of interaction with examples; the differences between experimental errors and interactions the experimental arrangements and analytical methods with intercrossing effects and their hybrid techniques.

5.1 Interaction Concepts and Judgments

5.1.1 The Concept of an Interaction

In some concrete experiments, not only do the various factors work, but also the factors that may work together, a function that is called interaction. Here's an example to illustrate the concept of interaction.

【Example 5.1.1】 Jiangsu Jianbi a project using 3 m diameter prestressed concrete pipe, steam curing in the production of the proposed to enhance the early strength, and mixed with rejection 0.5% DH-3 Qi water agent or 0.25% wood calcium super plasticizer. Mortar test conditions are as follows:

Jiangnan No. 600 oil well cement, cement : sand = 1 : 1.41, water-cement ratio of 0.35 * conducted a total of six tests, the results listed in Table 5.1.1.

Table 5.1.1

Experimental number	1	2	3	4	5	6
Kinds and admixtures of admixtures	0	DH-3 0.5%	MU Ca 0.25%	0	DH-30.5%	MU Ca 0.25%
Maintenance mode	Steam curing. After production, stop for 1 h, heat up for 3 h, and 70 ℃ Celsius for 2 h.			Standard maintenance 20 ±5 ℃		
Compressive strength (kg/cm^2)	370	410	35	300	364	212

Chapter 5 The Orthogonal Design of Interaction Function

These six trials can be divided into two orthogonal tables $L_4(2^3)$ to clarify why fh is the interaction between the two factors. How to judge whether there is interaction between two factors?

The first case examines the conservation mode A and additive agent B influence. The factors and levels are shown in Table 5.1.2. The test results and the calculation results are shown in Table 5.1.3.

Table 5.1.2 Factor level table

Level	Factor	
	A. Maintenance mode	B. Kinds and admixtures of admixtures
1	Standard maintenance	DH – 30.5%
2	Steam curing	Mu-Ca 0.25%

From the visual analysis we can see:
1. The most important thing is the extra-banging agent. DH – 3 is much better than wood calcium.
2. Steam Mangshi less than standard conservation. Suddenly, I did not agree with a rule.

Table 5.1.3 $L_4(2^3)$ Test results and calculation

Experimental number	A	B	A × B	Compressive strength (kg/cm²)
	1	2	3	
1	1 (Standard maintenance)	1 (DH-3)	1	364
2	2 (Steam curing)	1 (DH-3)	2	410
3	1 (Standard maintenance)	2 (Mu-Ca)	2	212
4	2 (Steam curing)	2 (Mu-Ca)	1	35
K_1	576	774	399	
K_2	445	247	622	
K_1	288	387	200	SUM:1 021
K_2	223	124	311	
R	65	263	111	

Here, the more important thing is that the curing methods and admixtures are well matched. It can not be said that the steam curing is not as good as the standard curing because the calcium and calcium are not suitable for the steam curing and the interaction between the calcium and steam curing. Thus, the mode of maintenance and admixture agents for total anti-tapenade total impact (i.e., the total effect of the test) is composed of each of them alone effect of the test plus the combination of two factors, This collocation is called the interaction between two factors. It is not hard to understand that the interaction of the two factors seems to be at work a "hypothetical factor" in addition to the separate effects of these two factors, but this is of no "horizontal" choice due to the vast expanse of its effect Depends on the first two factors and their level of collocation.

We write the interaction of factors A and B as $A \times B$. The order of the tests is: $B \to A \times B \to A$, that is, the interaction between additives and curing methods is more important than the curing effect alone.

Let us look at " under the conservation methods and blending do not doped. The factors and levels are shown in Table 5.1.4. The test results and the range calculation results are shown in Table 5.1.5.

Table 5.1.4 Factor level table

Level	Factor	
	A. Maintenance mode	B. DH – 3 volume
1	Standard maintenance	0
2	Steam curing	0.5%

Table 5.1.5 $L_4(2^3)$ Test results and calculation

Experimental number	A	B		Compressive strength (kg/cm^2)
	1	2	3	
1	1 (Standard maintenance)	1 (0)	1	300
2	2 (Steam curing)	1 (0)	2	370
3	1 (Standard maintenance)	2 (0.5)	2	364
4	2 (Steam curing)	2 (0.5)	1	410
K_1	664	670	710	
K_2	780	774	734	
K_1	332	335	355	SUM:1 444
K_2	390	387	367	
R	58	52	12	

From the visual analysis we can see:.

1. Steam curing is clearly better than standard curing.
2. Doping DH-3 is better than non-doping.
3. As a rule of thumb, the margins for empty columns are so small that within the limits of the general experimental error, there is no interaction between these two factors as a test error.

5.1.2 The Judgment of Interaction

There is a need for interaction between the two factors on the law two gods situation. Painted Figure 5.1.1. Figure 5.1.1 disabilities 1 saw there is interaction, the larger the difference between the two lines I did not. Hengshuai interaction, the two parallel rectangular body. Due to the existence of experimental errors, the two lines are not: may be completely parallel, only generally flat, we can determine the interaction is small or not.

Chapter 5 The Orthogonal Design of Interaction Function

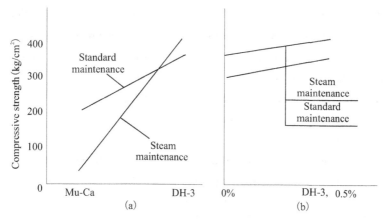

Figure 5.1.1

(a) The interaction of curing methods and agent species; (b) No interaction between curing mode and DH-3 dose

Sometimes due to the interference of experimental errors, it is difficult to judge from the drawing alone, and an analysis of variance is needed. Variance analysis gives the sum of squares of the variances caused by the interaction, to test for the significance of the interaction, and to be more accurate than the visual analysis for the edge. However, to use variance analysis to determine the interaction, there is no empty column as the test error estimate, there should be repeated test data provided in order to F test. As previously described, the work of retesting is greater than that of zinc and the significance of the interaction can be tested using data from repeated sampling. The F test for repeated data or data from samples is shown in Examples 3.2.3 and 3.2.4.

5.2 Differences between Interaction and Experimental Error

Look at an example first.

【Figure 5.2.1】 Nanjing Institute of Hydraulic Lee Research Institute of the two concrete super plasticizer MF and NNO line vibrator work to enhance the effect of the test. Vibrating process consists of two kinds of high-frequency plugging (Vibration frequency 10 560 times/min) junior high school frequency vibration (vibration frequency 2 800 times/min). The test results are shown in Table 5.2.1.

Table 5.2.1

No.	Dose(%)	Concrete mix ratio	Water cement ratio	Cement content (kg/m^3)	Water reduction rate(%)	Slump(cm)	7-day Compressive strength(kg/cm^2)	
							Taishine	Insert mode
0	0	1:1.75:3.4	0.55	358	0	3.3	190	202
N	NNO 0.8	1:1.75:3.4	0.48	367	10.7	3.0	254	267
M	MF0.8	1:1.75:3.4	0.40	375	20.2	4.0	196	313

Orthogonal Design in Concrete Application

The purpose of this test is to compare the enhancement effect of the super plasticizer MF and NNO on different conditions. Its assessment indicators for 7 - day compressive strength, the higher the intensity the better. We can use the table $L_4(2^3)$ divided into two cases into analysis more. Blank columns for unplanned factors can be used to estimate experimental errors or interactions. This example illustrates the difference between interaction and experimental error.

1. Develop factors for both scenarios in Table 5.2.2 and Table 5.2.3.
2. Determine the two scenarios test program.

Table $L_4(2^3)$ to arrange the test only' accounted for two columns vacated a column can be estimated experimental error or interaction. Take up a total of two tables 8 test.

Table 5.2.2 Factor level table

Level	Factor	
	A. NNO Dose %	B. Shine mode
1	0	Tai shine
2	0.8	Insert mode

Table 5.2.3 Factor level table

Level	Factor	
	A. MF Dose %	B. Shine mode
1	0	Tai shine
2	0.8	Insert mode

Benchmarks without water-reducing agent are common to both types of water-reducing admixtures and can save two tests, so the two tables can be completed in only six tests. Table 5.2.1 test results were filled in $L_4(2^3)$ to the right, in Table 5.2.4.

3. Analysis of test results for the first case.

Factor A said that the new A_1 mixed with NNO favorable, take the factor B shows insert vibration favorable, take B_1, the optimal combination is A_2B_2 that No. 4 test, $R_7 = 267$.

The order of the factors is: $A \to B \to$ the empty column, the empty column is the error of the experimental error estimate, indicating that the error of this test is very small.

For the second case:

Factor A shows that it is advantageous to mix MF and take A_2. It is worth knowing that the effect of MF and NNO is even worse than that of the first one. Factor B illustrates the vibration insertion benefit, as compared to factor B in the seventh case.

Chapter 5 The Orthogonal Design of Interaction Function

Table 5.2.4 $L_4(2^3)$ Test results and calculation

	The first case					The second case			
Experimental number	A. NNO dose	B. Vibration technology	$A \times B$	R_7	Experimental number	A. MF dose	B. Vibration technology	$A \times B$	R_7
	1	2	3			1	2	3	
1	1(0)	1(Platform vibration)	1	190	1	1(Platform vibration)	1	1	190
2	2(0.8)	1(Plugged in vibration)	2	254	2	2(0.8)	1(Plugged in vibration)	2	196
3	1(0)	2(Platform vibration)	2	202	3	1(0)	2(Platform vibration)	2	202
4	2(0.8)	2(Plugged in vibration)	1	267	4	2(0.8)	2(Plugged in vibration)	1	313
K_1	392	444	457	SUM: 913	K_1	392	336	503	SUM: 901
K_2	521	469	456		K_2	509	515	398	
R	129	25	1		R	117	129	105	

The effect is much better. The primary and secondary elements of the sequence is P empty. Empty here. The difference between the columns is very similar to that of column B, apparently not due to trial error, which is much larger than the empty column of the first case. In this case, the role of vibration insertion can be combined with the role of MF and work together, that is, A, while B has interaction. The role of plug-in vibration and can not be combined with and play a joint role. So the empty case in the first case is an estimate of the experimental error. Judging from the professional knowledge, MF is an aerated g water reducer. Although its water-reducing rate is much larger than that of NNO due to aeration, its enhancement effect is less than that of NNO under the conditions of Taichung. In the collar Zhen conditions, due to mechanical deforming, so that the enhanced effect of MF greatly improved. The optimal combination of the second case is $A_2 B_2$. The main difference from the first case is the interaction between A and B, $A \times B$.

This example tells us that the use of empty columns to estimate experimental error is made with little or no negligible interaction between elements. If the interaction is significant, it is impractical to estimate the experimental error using the null column. In the coagulation Interaction does not occur often in experiments. It can be empirically estimated using empty columns in some special cases, the question of "how to deal with interactions" and experimental errors will be discussed below.

5.3 The Experimental Arrangements and Analytical Methods with Interaction Function

The concept of interaction and its intuitive judgment as well as its difference from the test error are introduced above. The following examples and examples further illustrate the experiment arrangement and analysis method with interaction.

【Example 5.3.1】 with two different steam curing time and vibrating mixed soil reinforcement effect of the comparative test. The factors and levels in the experiment are listed in Table 5.3.1, and the influence of test domes A, B, C and $A \times B$, $A \times C$, The process conditions.

Table 5.3.1 Factor level table

Level	Factor		
	A. Water cement ratio	B. Shine mode	C. Maintenance mode(h)
1	0.40	Tai shine	3
2	0.45	Insert mode	4

1. Experimental arrangements.

a) Table design, as A, B, C are two levels, in addition to A, B, C, but also examine the effects of $A \times B$, $A \times C$, $B \times C$, because: This selection of orthogonal table $L_8(2^7)$ Suitable.

The principle of the header design is:

a. When you know the interaction between the factors: you need the interaction table (or the interaction of the header design).

b. So that the main effects and experimental errors and interactions have a "mixed phenomenon." There is a table in the Appendix m (Orthogonal Table) called Inter-column 1: Action Table (see Table 5.3.2). This type of table is very important for how to arrange the test. Table 5.3.2 illustrates the design method of this example.

Chapter 5 The Orthogonal Design of Interaction Function

Table 5.3.2 $L_8(2^7)$ Intercolumn interaction table

No.	No.					
	1	2	3	4	5	6
7	6	5	4	3	2	1
6	7	4	5	2	3	
5	4	7	6	1		
4	5	6	7			
3	2	1				
2	3					

If A is placed in the first column, B on the second column, look-up Table 5.3.2 "1" column "2" line, corresponding to 3, that is, the third column reflects the $A \times B$, if A on Column 3, B on the 4th line, look-up Table 5.3.2, corresponding to the "3" column "4" The number of rows is 7, that $A \times B$ in column 7. Now put A in the orthogonal table in the first Column B on the orthogonal table in the second column, according to the above principle, then C can not be placed in the orthogonal table in the third column, or C and $A \times B$ mixed together to produce this phenomenon is called mixed. So C is placed in column 4 and $A \times C$ is found in Table 5.3.2 (column 1 and row 4 are placed in column 5 of the orthogonal table. Column 6 is for column 7 of $B \times C$ (column 2 and 4) Empty, as a test error estimates ten. The header design:

No.	1	2	3	4	5	6	7
Factor	A	B	$A \times B$	C	$A \times C$	$B \times C$	

For convenience, appendix III Table 1, after the interaction between the two columns are listed in this table header design or interaction column comments, for readers to directly check.

b. Test scheme: from the head design, the test scheme is given by the 1, 2, 4 three column, the interaction can be analyzed through the three column of the three column, and the seventh column is used as the estimate of the test error. Test scheme and test result column and Table 5.3.3.

Table 5.3.3 Test results and calculation

Experimental number	A	B	$A \times B$	C	$A \times C$	$B \times C$		7-day Compressive strength (kg/cm^2)
	1	2	3	4	5	6	7	
1	1(0.40)	1(Plugged in vibration)	1	1	1	1	1	169
2	1	1	1	2	2	2	2	178
3	1	2(vibration)	2	1	1	2	2	173
4	1	2	2	2	2	1	1	272
5	2(0.45)	1	2	1	2	1	2	146
6	2	1	2	2	1	2	1	169
7	2	2	1	1	2	2	1	194
8	2	2	1	2	1	1	2	215
K_1	892	662	662	782	826	802	804	
K_2	724	954	860	834	790	814	812	SUM:1 618
R	168	292	104	52	36	12	8	

Orthogonal Design in Concrete Application

2. Analysis of the test results

Compressive strength and its range of 7 days of 8 tests are listed in the table
a) Right and below Table 5.3.3.
a. Visual Analysis Derived from Table 5.3.3. $B > A > A \times B > C > A \times C > B \times C$
b. According to the size of the factors K value, take A is A_1, take B to B is B_2, take C to C_2 as well. In general, to obtain a higher intensity of 7 days, the combination of conditions can be set as $A_1B_2C_2$.
c. Since $A \times B$ has an impact on R_7, the best level for selecting A and B is also subject to the $A \times B$ level. Therefore, further interaction analysis is needed.

The intuitive analysis of the interaction between A and B is: Add the values of R_7, corresponding to the same level of B, divided by the number of additions, that is, the K value of the interaction at a certain level of $A \times B$. The calculation results are shown in Table 5.3.4.

Table 5.3.4

B	A	
	A_1	A_2
B_1	173.5	157.5
B_2	272.5	204.5

Seen from the above table, A_1B_2 combination can get higher 7 d compressive strength, the value of 272.5 kg/cm^2. Consistent the above-identified process conditions $A_1B_2C_2$. Therefore, after considering the interaction, the optimal combination condition is still $A_1B_2C_2$ the fourth experimental condition.

d. Since $A \times C$ and $B \times C$ have a small influence on R_7, they can be combined with the seventh column to jointly estimate the experimental error.

In general, the end of the analysis can terminate the analysis. Because the optimum process conditions derived from analysis is $A_1B_2C_2$, which is the fourth test, the 7-day compressive strength of 272 kg/cm^2 is also a high test 8.

Explain the intuitive analysis is correct. However, the strength of No. 3 test was the same as that of No. 4 test (R_7 = 273 kg/cm^2) of No. 3 test, and the curing time of No. 3 test was shorter by 1 hour than that of No. 4 test, and the condition of No. 3 test was good, ANOVA needs to be done to learn more.

b) Analysis of variance

For two-level factors, the calculation of the sum of squares has a simpler formula

$$S_i = \frac{(K_1 - K_2)^2}{n}$$

Where n one by one the number of tests.

According to the data in Table 5.3.3, for example, the first column of the side and according to common calculation method:

$$P = \frac{1}{8} \times 1\,616^2 = 326\,432$$

$$Q_A = \frac{1}{4} \times (892^2 + 724^2) = 329\,960$$

$$S_A = Q_A - P = 329\,960 - 326\,432 = 3\,528$$

By simplifying the calculation formula

Chapter 5 The Orthogonal Design of Interaction Function

$$S_A = \frac{1}{n}(K_1 - K_2)^2 = 168^2/8 = 3\,528$$

This method is applicable to any two level factors. The other columns are calculated in the first column. Each column has a degree of freedom of one.

An analysis of variance is given in Table 5.3.5.

Analysis of variance showed that A and B had a particularly significant effect on the 7-day intensity; $A \times B$ had a significant effect, while the volume C had an effect on the intensity. Therefore, A and B should be set to have a large value of $A_1 B_2$, while the level of C can be arbitrarily chosen, that is, the best conditions for recording are excellent. Therefore, both No. 3 and No. 4 tests can be used for the optimal process conditions 7 and the test results are exactly the same.

Table 5.3.5 Variance analysis table

Variation source	Sum of squares	Freedom	Mean square	F value	Critical value
A	$S_A = 3\,528$	1	3 528	56.3	$F_{0.01}(1,3) = 34.1$
B	$S_B = 10\,658$	1	10 658	170.0	$F_{0.05}(1,3) = 10.1$
$A \times B$	$S_{A \times B} = 1\,352$	1	1 352	21.6	$F_{0.10}(1,3) = 5.5$
C	$S_c = 338$	1	338	5.4	
$A \times C$	$S_{A \times C} = 1\,352$	1			
$B \times C$	$S_{B \times C} = 1\,352$	1			
Empty column	$S_E = 1\,352$	1			
Error	$S_e = 0.398\,1$	3	62.7		

Taking into account the reduction of 1 hour combination of curing time, the final selection more realistic.

From this example we see that the projection design not only analyzes the error r but also analyzes the interaction, which is the advantage of the orthogonal design m. In the mean time, the analysis of variance in the orthogonal k not only separates each factor Squares and handcuffed calculations and calculation of the square root of the sudden interaction are also the same specifications. The reader may have noticed that this example is a full-blown three-factor cross-sectional test, which is much more annoying if calculated using the cross-grouped formula.

【Example 5.3.2】 Research Institute of Science and Technology of Communications and Ministry of Water Resources Survey and Design Amines In the rapid test of concrete strength, factors influencing the quick-hardening strength of cement mortar were investigated. Four factors were chosen for the experiment, with each factor taking two levels (see Table 5.3.6). The purpose of the experiment is to investigate the influence of various factors on the strength of fast-hardening mortar, especially the interaction between factors, so as to find out the basic rules of quick-hardening mortar.

Assessment indicators: a size of 3.16 cm × 3.16 cm × 5 cm mortar strong skin specimens. Requires not only save the gray than 1.4 when the intensity of business, but also in the gray water than higher one.

Table 5.3.6 Factor level table

Level	Factor			
	A. Maintenance mode	B. Coagulant	C. Water cement ratio	D. Cement
1	Autoclave 1.5 h	N	1.4	Peking 425 cement
2	70~80 ℃	Y	2.5	Changbaishan 425 cement

Orthogonal Design in Concrete Application

1. Test Arrangements

a) Header Design

In Example 5.3.1, we introduced a header design method for the interaction of two factors. We refer to the interaction between two factors (for example, $A \times B$) as the first-order interaction. When there are interactions between three or more factors, we call it the second-order interaction. For example, $A \times B \times C$ is a second-order interaction. Such interaction can also be calculated using a certain column of the orthogonal table. In this example, besides examining the main effects of B, C, D and their In addition to the primary interactions $A \times B$, $A \times C$, $B \times C$, $A \times D$, $B \times D$, and $C \times D$, there is also a discussion of the existence of secondary interactions $A \times B \times C$ (in the general case, secondary interactions are neglected.) For the four-factor two-level trial If only the main effects and their primary interactions are examined, the header design $L_{16}(2^{15})$ in Table 1 of Appendix III can be directly investigated:

No.	1	2	3	4	5	6	7	8	9	10	11	12	13	14	15
Factor	A	B	$A \times B$	C	$A \times C$	$B \times C$		D	$A \times D$	$B \times D$		$C \times D$			

Now to examine the secondary interaction $A \times B \times C$, which column it arranged. Similar to arranging the first-level interaction, the interaction table of $L_{16}(2^{15})$, (see the appendix III in table 1) can be used. The corresponding number of "3" column "4" That is, $A \times B \times C$ is in column 7.

b) Experimental program

Seen from the above header design, given by the 1st, 2nd, 4th and 8th columns, the first-level interaction is investigated in the third, sixth, sixth, sixth, ninth, tenth and sixth columns; Level 2 interaction; column 1, 13, 14, 15 four columns as an estimation of experimental error. The test plan is shown in Table 5.3.7.

2. The analysis of test results

The results of the visual analysis of the 16 tests and the sum of squares of the columns are given below in Table 5.3.7.

a) Obtained from the very poor size of the various factors and their interaction on the mortar strength of the order of order:

$$B \rightarrow A \rightarrow A \times B \cdot C \rightarrow A \times C \rightarrow B \times C \rightarrow A \times B \times C \rightarrow D \rightarrow BD \rightarrow CD$$

b) ANOVA results are shown in Table 5.3.8. It can be seen that the effects of main effects A, B, C, D and interactions $A \times B$, $A \times C$, $B \times C$ and $A \times B \times C$ on the strength of mortar have a significant impact on $B \times D$ and $C \times D$.

The effect of $A \times D$ is small and can be combined with empty columns to jointly estimate the experimental error with a value of $\hat{\sigma}_e = \sqrt{22.0628} = 4.7 \text{ kg/cm}^2$.

c) Because the interaction between the factors is particularly significant or significant, the selection of A, B, C, D and their internal rules are rooted, and they are chosen and explained according to their collocation.

Similar to the first-level interaction, the visual comparison of A with B and C is shown in Table 5.3.9. Visual comparison from the $A \times B \times C$ seen, Take $A_1 B_2 C_1$ and $A_1 B_2 C_2$ as good

That is, the technology can effectively reflect the impact of different sizes of gray water.

Table 5.3.7 $L_{16}(2^{15})$ Test results and calculation

Experimental number	A 1	B 2	A×B 3	C 4	A×C 5	B×C 6	A×B×C 7	D 8	A×D 9	B×D 10	 11	C×D 12	 13	 14	 15	Compressive strength (kg/cm²)
1	1	1	1	1	1	1	1	1	1	1	1	1	1	1	1	3
2	1	1	1	1	1	1	1	2	2	2	2	2	2	2	2	1
3	1	1	1	2	2	2	2	1	1	1	1	2	2	2	2	30
4	1	1	1	2	2	2	2	2	2	2	2	1	1	1	1	19
5	1	2	2	1	1	2	2	1	1	2	2	1	1	2	2	57
6	1	2	2	1	1	2	2	2	2	1	1	2	2	1	1	42
7	1	2	2	2	2	1	1	1	1	2	2	2	2	1	1	196
8	1	2	2	2	2	1	1	2	2	1	1	1	1	2	2	168
9	2	1	2	1	2	1	2	1	2	1	2	1	2	1	2	2
10	2	1	2	1	2	1	2	2	1	2	1	2	1	2	1	1
11	2	1	2	2	1	2	1	1	2	1	2	2	1	2	1	3
12	2	1	2	2	1	2	1	2	1	2	1	1	2	1	2	2
13	2	2	1	1	2	2	1	1	2	2	1	1	2	2	1	14
14	2	2	1	1	2	2	1	2	1	1	2	2	1	1	2	7
15	2	2	1	2	1	1	2	1	2	2	1	2	1	1	2	43
16	2	2	1	2	1	1	2	2	1	1	2	1	2	2	1	9
K_1	516	61	126	160	423	127	394	348	305	264	274	301	287	303	287	SUM:597
K_2	81	536	471	437	174	470	203	249	292	333	323	296	310	294	310	
R	435	475	345	277	249	343	191	99	13	69	49	5	23	9	23	
S_i	11 826.562 5	14 101.562 5	7 439.062 5	4 795.562 5	3 875.062 5	7 353.062 5	2 280.062 5	612.562 5	297.562 5	297.562 5	150.062 5	1.562 5	60.062 5	5.062 5	33.062 5	

· 87 ·

Table 5.3.8 Variance analysis table

Variation source	Sum of squares	Freedom	Mean square	F value	Critical value
A	11 826.562 5	1	11 826.562 5	536.05	$F_{0.01}(1,5) = 16.3$
B	14 101.062 5	1	14 101.062 5	639.16	$F_{0.05}(1,5) = 6.6$
$A \times B$	7 439.062 5	1	7 439.062 5	337.18	
C	7 353.062 5	1	7 353.062 5	333.28	
$A \times C$	4 795.560 2	1	4 795.560 2	217.35	
$B \times C$	3 875.062 5	1	3 875.062 5	175.64	
$A \times B \times C$	2 280.062 5	1	2 280.062 5	103.34	
D	612.562 5	1	612.562 5	27.76	
$B \times D$	297.562 5	1	297.562 5	13.49	
$C \times D$	150.262 5	1	150.262 5	6.80	
Errore	110.312 5	5	22.062 5		
SUM	52 841.437 5	15			

Table 5.3.9

		A_1	A_2
B_1	C_1	$(3+1)/2 = 2$	$(2+1)/2 = 1.5$
	C_2	$(30+19)/2 = 24.5$	$(3+2)/2 = 2.5$
B_2	C_1	$(57+42)/2 = 49.5$	$(14+7)/2 = 10.5$
	C_2	$(196+168)/2 = 182$	$(43+9)/2 = 26$

Visual comparison of B and D is shown in Table 5.3.10.

The visual comparison of $B \times D$ shows that it is good, that is, accelerating accelerator is effective for both types of cement and is more effective for ordinary cement.

Visual comparison of C and D is shown in Table 5.3.11.

Table 5.3.10

	B_1	B_2
D_1	$(3+30+2+3)/4 = 9.5$	$(57+196+14+43)/4 = 77.5$
D_2	$(1+19+2+8)/4 = 6.25$	$(42+168+7+9)/4 = 56.5$

Table 5.3.11

	C_1	C_2
D_1	$(3+57+2+14)/4 = 19$	$(30+196+3+43)/4 = 68$
D_2	$(1+42+1+7)/4 = 12.75$	$(19+168+2+9)/4 = 49.5$

According to the comparison of $C \times D$, it is better to use C_2D_1 and C_2D_2, that is, the cement type and the gray water ratio work together, and the common cement and gray water ratio work together better.

Chapter 5 The Orthogonal Design of Interaction Function

With professional knowledge analysis, factors B_2 (dedicated to promote coagulation) the most important. The importance of factor B (pressure steaming method) is second only to Burgundy coagulant and steam curing combination $A \times B$, its effect and the result of C (gray-water ratio). In addition, steam curing C, plus (B_2) interacts with the C ratio ($A \times C$, $B \times C$), and both of them interact with the C ratio ($A \times B \times C$) Can show different potential strength in a short period of time. It can also be said that these laws and regulations constitute the basic principle of "pro-coagulation and pressure steaming technology", and can be applied to common cement and slag cement.

In practice, the factors C, D are often not arbitrarily chosen, both gray-water ratio and cement varieties are likely to encounter. If it is determined $A_1 B_2$ to be the heaviest.

In fact, only four tests of test numbers 5, 6, 7, and 8 are useful in the 16 tests in Table 5.3.7, including two gray-water ratios and two different cement types. Thus, it is not only the number of tests that is used to design this interactive method of analysis, but also the analysis results are far more profound than the traditional methods of analysis.

3. discussion

In this case, if we consider $A \times B \times C$, which is the secondary interaction between factors, the seventh column is empty column, which may be considered a false column, and the 191 column is a very small value, which is extremely bad in each column, If it is incorporated in the error, not only the experimental error is significantly increased, but also the F test of factors D, $B \times D$ and $C \times D$ is not significant, which distorts the analysis result. Therefore, under the condition that the control of a test condition is relatively strict, when the empty column is very large, the interaction must be considered first, or it is not known for a moment whether there is an interaction or not. Under the circumstances, the header should be designed so as not to confound the interaction, which is especially noticeable when revealing internal rules. However, because of the secondary interaction between purple, it is extremely rare or negligible in concrete experiments, and in some cases it can only be interactive. The effects are mixed and the promiscuous techniques are discussed in the next section.

5.4 Mixed Skills

In the multi-factorial experiment, the main advantage of orthogonal design is that it can save the number of experiments. However, this advantage is greatly diminished if more analytic interactions are required. Example 5.3.1 test, due to the analysis of the three interactions, in the orthogonal table Jing account for three columns, and these three numbers can not go to the other row due to abandon, otherwise it will produce a "mixed phenomenon." If you examine the factors and more analysis of the interaction also requires more, the required number of tests must be many, which in the concrete material test.

It is a contradiction that may be encountered. To solve this contradiction, we must analyze the role of interaction in orthogonal design, that is, hybrid techniques. The so-called hybrid technique is to reduce the number of trials, using smaller orthogonal tables to arrange more factors, knowing that there are mixed and deliberately let them mixed approach.

Orthogonal Design in Concrete Application

If the main purpose of the test is to find the internal rules of the test, the work S and the expense of the test are not too large. In this case, the orthogonal test can be chosen to avoid mixing if the main purpose is to find a better the process conditions, the objective conditions are not allowed to do too much testing, and then choose a smaller orthogonal table. Due to the selection of a smaller orthogonal table, inevitably have to produce mixed phenomenon, whether it can achieve the effect of period? This is illustrated by Example 5.3.1.

Example 5.3.1 if only to select the better process conditions, and the objective conditions do not allow 8 tests, then had to choose orthogonal table that has three columns, the interaction of any two columns The role of 3 columns, that is, if the A and B were placed in the first column and the second column, $A \times B$ in column 3. Now turn A, B, C respectively, on the 1, 2, 3 columns, the header design

No.	1	2	3
Factor	A $B \times C$	B $A \times C$	C $A \times B$

Can be seen in the three columns are mixed. $L_4(2^3)$ arrangement of the four tests was $L_8(2^7)$ arranged on the 1st, the 4th, the 6th and the 7th test, the test arrangement and results are shown in Table 5.4.1.

From Table 5.4.1 K value size obtained A_1 take A_1, B take B_1 C_1 take C better, the better process conditions that the four tests in the No. 2 test (ie, $L_8(2^7)$ No. 4 test). Because each column is mixed, it is not possible to analyze the order of major and minor influences on the 7-day intensities due to and interactions and their effect is significant. Although smaller orthonormal are used. However, due to the fact that the orthogonal tables used the principle of equilibrium dispersion in arranging the test, the better process conditions selected by 2 s are basically the same as the results of $L_8(2^7)$ analysis.

Table 5.4.1

Experimental number	A $B \times C$ 1	B $A \times C$ 2	C $A \times B$ 3	7-day Compressive strength(kg/cm^2)
1	1(0.4)	1(Plugged in vibration)	1(3 h)	169
2	1(0.4)	2(vibration)	2(4 h)	272
3	2(0.45)	1(Plugged in vibration)	2(4 h)	169
4	2(0.45)	2(vibration)	1(3 h)	194
K_1	441	338	363	SUM:804
K_2	363	466	441	

Thus, $L_4(2^3)$ is inferior to $L_8(2^7)$ in knowing the internal law of matter, but the former is reduced by half (and in most cases, more) than the latter. Therefore, when there are more factors due to more investigations and objective conditions may not allow more experiments, in order to reduce the number of experiments, some or all of the hybrid methods may be used to arrange experiments on smaller orthogonal tables, that is, With or without interaction, through the test to remove minor causes, narrow the scope of the test, and then explore the internal laws of things.

Chapter 6
Regression Analysis of Orthogonal Design

The so – called regression analysis of orthogonal design is the use of the "orthogonally" of orthogonal design, on the one hand, and the two principles of "decentralization" and Planned, purposeful, Orthogonal tables: Less earthquakes are scheduled for fewer tests to & better experimental results; on the other hand, the least-squares method of multiplication is used to derive data from measured data The important reasons for the establishment of empirical formulas between examinations and examinations, and the merits of both, are combined organically, thus forming a back-analysis of orthodoxy design.

In the concrete experiment, there is a correlation between some parameters of the normal pressure compounding ratio and some characteristic standards of concrete or between various characteristic indexes. For example, ① the relation between the compressive elasticity and the ratio of gray water to water, ② the use of water and gray water than the lack of relationship between; ③ tensile strength and compressive strength between the relationship and so on. Mathematical statistics to deal with similar ①, ② of the problem, known as regression analysis; to deal with similar ③ of the problem, called the correlation analysis. Regression analysis and correlation analysis is a mathematical tool to study and analyze the relationship between change and change. Analysis and related analysis, commonly known as regression analysis.

6.1 Introduction of Regression Analysis

6.1.1 Linear Regression Equation with One Element

We use a linear regression to introduce the basic method of regression analysis.

The univariate linear regression analysis is a problem we often encounter with the so-called straight line, that is, the two variables z and have a certain relationship. Through experiments, we get the data and find the empirical formula between the two. Look at an example.

【Example 6.1.1】 The column a under the conditions of a total check, get the concrete 28-day compressive strength and gray water ratio x data are shown in Table 6.1.1. Find the relationship between intensity y and gray-water ratio.

Orthogonal Design in Concrete Application

Table 6.1.1   The relationship between the compressive strength and the water cement ratio

Water cement ratio	0.40	0.45	0.50	0.55	0.60	0.65	0.70	0.80
cement Water ratio x	2.500	2.222	2.000	1.818	1.667	1.538	1.429	1.333
28-day Compressive strength y (kg/cm^2)	363	353	282	240	230	205	184	150

Draw the data points on the coordinate paper, as shown in Figure 6.1.1. As can be seen from Figure 6.1.1, the data points are near a straight line and the stem is represented by a straight-line equation

$$\hat{y} = a + bx \tag{6.1.1}$$

It is called y regression line for x how to determine this regression line and its equation? First you need to give the formula.

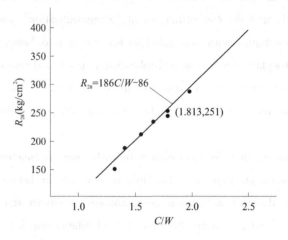

Figure 6.1.1 *The relationship between the compressive strength and C/W*

Coincidentally: Mutual cross dating of the measured value (x_i, y_i) $i = 1, 2, \cdots, n$, From their point of view ± Shy a line relationship, this relationship can be expressed as

$$\hat{y}_i = \alpha + \beta x_i + \varepsilon_i, \quad i = 1, 2, \cdots, n \tag{6.1.2}$$

α and β is constant. ε_i is experimental error. If ignore the error difference, y and x is a strict linear relationship: The y value of the corresponding x_i is recorded as $\hat{y}_i$

$$\hat{y}_i = \alpha + \beta \hat{x}_i, \quad i = 1, 2, \cdots, n \tag{6.1.3}$$

Since α and β is generally unknown, how to estimate α and β after getting n observations? A common method is to select α and β and to minimize $\sum_{i=1}^{n} (y_i - \hat{y}_i)^2$. This is the least squares principle. According to this principle, and estimates were expressed as a and b

$$b = \hat{\beta} = \frac{\sum_{i=1}^{n}(x_i - \bar{x})(y_i - \bar{y})}{\sum_{i=1}^{n}(x_i - \bar{x})^2} \tag{6.1.4}$$

$$a = \hat{a} = \bar{y} - b\bar{x} \tag{6.1.5}$$

in which,

Chapter 6 Regression Analysis of Orthogonal Design

$$\bar{x} = \frac{1}{n}\sum_{i=1}^{n} x_i$$

$$\bar{y} = \frac{1}{n}\sum_{i=1}^{n} y_i$$

For the convenience of narration

$$L_{xx} = \sum_{i=1}^{n}(x_i - \bar{x})^2 = \sum_{i=1}^{n} x_i^2 - \frac{(\sum_{i=1}^{n} x_i^2)^2}{n} \tag{6.1.6}$$

$$L_{xy} = \sum_{i=1}^{n} x_i y_i - \frac{1}{n}(\sum_{i=1}^{n} x_i)(\sum_{i=1}^{n} y_i) \tag{6.1.7}$$

$$L_{yy} = \sum_{i=1}^{n}(y_i - \bar{y})^2 = \sum_{i=1}^{n} y_i^2 - \frac{(\sum_{i=1}^{n} y_i^2)^2}{n} \tag{6.1.8}$$

The calculation time is listed in Table 6.1.2.

Table 6.1.2 One element regression calculation table

No.	x	y	x^2	y^2	xy
1	2.500	363	6.250	131 800	907.5
2	2.222	353	4.937	124 600	784.4
3	2.000	282	4.000	79 520	564.0
4	1.818	240	3.306	57 600	436.3
5	1.667	230	2.779	52 900	383.3
6	1.538	206	2.366	42 440	346.8
7	1.429	184	2.042	33 860	262.9
8	1.333	150	1.777	22 500	200.0
Σ	14.507	2 008	27.457	545 220	3 855.3

The last line of Table 6.1.2 is the sum of the columns, with the amount we need to calculate this line:

$$\sum x = 14.507, \quad \sum y = 2\,008, \quad n = 8$$

$$\bar{x} = 1.813, \quad \bar{y} = 251, \quad \sum xy = 3\,855.3$$

$$\sum x^2 = 27.457, \quad \sum y^2 = 545\,220, \quad (\sum x)(\sum y)/n = 3\,641.3$$

$$(\sum x)^2/n = 26.307, \quad (\sum y)^2/n = 5\,040\,008$$

$$L_{xx} = \sum_{i=1}^{n} x_i^2 - \frac{(\sum_{i=1}^{n} x_i^2)^2}{n} = 1.15, \quad L_{yy} = \sum_{i=1}^{n} y_i^2 - \frac{(\sum_{i=1}^{n} y_i^2)^2}{n} = 41\,212$$

$$L_{xy} = \sum_{i=1}^{n} x_i y_i - \frac{1}{n}(\sum_{i=1}^{n} x_i)(\sum_{i=1}^{n} y_i) = 214$$

So:

$$b = \frac{L_{xy}}{L_{xx}} = 186$$

$$a = \bar{y} - b\bar{x} = 251 - 186 \times 1.813 = -86$$

$$\bar{y} = 186x - 86$$

The above equation is the regression equation. This regression equation goes by $(\bar{x}, \bar{y})$ at this point. Then let x take a number of x_0, substituted into the regression equation to find the corresponding y_0, the link $(\bar{x}, \bar{y})$ and (x_0, y_0) two points, is the regression line, see Figure 6.1.1.

The parameter b is called the regression coefficient, and its physical meaning is: when the gray water ratio increases or decreases 0.1, the compressive strength increases or decreases by 28.6 kg/cm² in 28-day.

6.1.2 The Linear Correlation Coefficient r and its Test

The linear correlation coefficient, r, shows the closeness of 1 and linearity. Its physical meaning is shown in Figure 6.1.2. The closer the absolute value of r is to 1, the better the linear correlation between y and x. If it is close to 0, it means that there is no linear relationship between y and x.

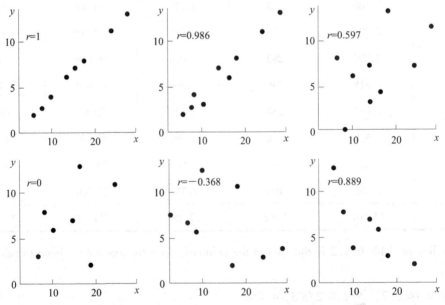

Figure 6.1.2 *Correlation Coefficient scatter diagram*

The correlation coefficient is defined as follows:

$$r = \frac{L_{xy}}{\sqrt{L_{xx}L_{yy}}} \tag{6.1.9}$$

Substitute the data of Example 6.1.1 into Equation 6.1.9

$$r = \frac{214}{\sqrt{1.15 \times 41\,212}} = 0.98$$

Describe a certain type of cement and labeling, concrete compressive strength and gray water ratio. When the sample n size is very small, the coefficient of variation of the bismuth-like coefficient of the

Chapter 6 Regression Analysis of Orthogonal Design

sample and the sample of the sample in the sample have large variability. This choice is no more than the minimum value, or test the overall correlation coefficient $\rho = 0$ is credible? In Appendix III Table 5 is given the correlation coefficient check table, the value of the table is the minimum value of the relationship between the value of the key, the correlation coefficient r found on the table on the value of the husband to test at a straight line to describe correlation is intentional. It is meaningful to consider the correlation between y and x in line. Example 6.1.1, $n = 8$, check appendix III Table 5 corresponds to a line with $n - 2 = 6$, with a corresponding number of 0.707 (confidence $a = 5\%$), thus $r = 0.98 > 0.707$, so the fitted line makes sense.

6.1.3 The Tropism of Precision

Because of the correlation between y and x, we know the value of a; we can not know the exact value of $\hat{y}$ but we can find the average of y from the regression equation. How long can be dragon? That is, with the measured point distribution into a straight line how accurate? Under the same x the measured value of y fluctuates according to the normal distribution. If the mean square deviation of the fluctuations can be calculated, the accuracy of the regression line can be estimated. The mean square deviation here is called the residual mean square deviation, indicating the precision of wiring. It is defined by the following formula:

$$S = \sqrt{\frac{1}{n-2} \sum_{i=1}^{n} (y_i - \hat{y})^2} \tag{6.1.10}$$

Inconvenient, switch to use

$$S = \sqrt{\frac{L_{yy} - L_{xy} \cdot b}{n - 2}} \tag{6.1.11}$$

Into the specific data, get:

$$S = \sqrt{\frac{41\,212 - 214 \times 86}{8 - 2}} = 15.3 \text{ kg/cm}^2$$

When using the regression equation to predict concrete strength, the chance of 95% will not be exceeded $2S = 30.6 \text{ kg/cm}^2$.

6.1.4 Analysis of Variance

In bivariate regression, however, the goal of variance squaring and r testing is the same. Similar to the one-way analysis of variance in Chapter Three, the variances $y_1, y_2, \cdots, y_n$ squads of variation squared and $S_T(L_{yy})$ are two partially due to the change of y caused by the change of x, and the other part caused by the difference of test I. The former is called the sum of squares of regression, denoted as S_N. The latter is called the residual square sum, denoted as S_e. then

$$S_T = S_N + S_e \tag{6.1.12}$$

can be calculated by:

$$S_N = bL_{xy} \tag{6.1.13}$$

$$S_e = L_{yy} - bL_{xy} \tag{6.1.14}$$

The corresponding degree of freedom

$f_N = 1, f_e = n - 2$

Take the value in

$S_N = 39\ 804$, $S_e = 1\ 408$, $S_T = 41\ 212$

Variance analysis column and table 6.1.3. The analysis of variance shows that the regression is particularly significant, indicating that there is a very close linear relationship between y(28-day pit pressure) and x work(gray water ratio). And r test conclusion is exactly the same. For a one-way regression analysis, generally do not need to make a variance and only test the correlation coefficient r.

Table 6.1.3 One element regression calculation variance analysis table

Variation source	Sum of squares	Freedom	Mean square	F value	Critical value
Regression	$S_R = 39\ 804$	1	39 804	169.4	$F_{0.01}(1,6) = 13.17$
Error	$S_e = 1\ 408$	6	235		
SUM	$S_T = 41\ 212$	7			

6.1.5   The Return of Money & Line Regression for the Money

Fitting straight lines is the most common and commonly used test data. Grasp the basic analysis of the above linear regression method can also handle some of the nonlinear regression problems, such as cement hydration heat or concrete adiabatic temperature and age relationship; compressive elastic modulus and age relationship; compression Intensity and anti-i-intensity relationship, etc., the solution is through the appropriate transformation of variables, the nonlinear regression is reduced to linear regression.

6.1.5.1   Hyperbolic Relationship

The adiabatic temperature rise of concrete, T_r or the hydration heat of cement, Q, is related to the relationship between age t and hyperbolic relation:

$$\left. \begin{array}{l} T_r = \dfrac{Mt}{N+t} \\ Q = \dfrac{Mt}{N+t} \end{array} \right\}$$

in which T_r = Adiabatic temperature rise of concrete, ℃

Q = Cement hydration heat, card / gram

T = age term, day

M, N = constant

The above formula shows when $t \to \infty$, T_r or $Q \to M$, M, Q Heat temperature or cement final hydration heat; when T or Q is equal to $\dfrac{M}{Z}$, then $N = t$, N is the number of days required for adiabatic or hydration development to reach half of the final value, characterizing its rate of development.

The above formula can be rewritten as

$$\left.\begin{aligned} \frac{t}{T_r} &= \frac{N}{M} + \frac{1}{M}t \\ \frac{t}{Q_r} &= \frac{N}{M} + \frac{1}{M}t \end{aligned}\right\}$$

let $\dfrac{t}{T_r} = y = \dfrac{t}{Q}$, $t = x$, then

$$y = \frac{N}{M} + \frac{1}{M}x$$

we can see $\dfrac{1}{M}$ is the slope of a straight line, while $\dfrac{N}{M}$ is the intercept of a straight line on the y-axis.

6.1.5.2 Exponential Relationship

The relationship between the instantaneous elastic modulus E and age t of concrete is approximate exponential formula

$$E = E_0(1 - \beta e^{-rt})$$

in which, E_0 = The final modulus of elasticity, kg/cm^2, extrapolated according to the test data obtained

β, r = Constant, the value depends on the mix and hardening conditions

Transform accounted for the type

$$1 - \frac{E}{E_0} = \beta e^{-rt}$$

Take a logarithm on both sides

$$\ln(1 - \frac{E}{E_0}) = \ln\beta - rt$$

When $y = \ln(1 - \dfrac{E}{E_0})$, $a = \ln\beta$, $x = t$, then

$$y = a - rx$$

Concrete tensile strength R and compressive strength R_y & approximate the relationship between the following formula

$$R_L = KR_y^m$$

in which k, m = constant Log on both sides

$$\lg R_L = \lg K + m\lg R_y$$
$$y = \lg R_L, a = \lg K, x = \lg R_y$$
$$y = a + mx$$

Finally, it should be pointed out that the linear regression we are discussing is made under the following assumptions:

1. when valve ξ(water cement ratio) fixsa value such as compressive strength η is a random variable with normal distribution when a variable.

2. The variance η_i has nothing to do with σ^2 and i. that is Constant to table Show.
3. The η_i parent population mean $E(\eta_0/\xi = x)$ is a linear function of $x(\beta \neq 0)$ $E(\eta/\xi = x) = a + \beta x$.
4. Each pair of observations (x_i, y_i) and the other pair (x_i, y_i) are observed independently of each other in probability.

In the processing of concrete test data, the above assumptions are generally satisfied. Under certain conditions, it is difficult to know whether a certain assumption complies with the actual situation, which can be statistically inferred by the method of hypothesis testing. The scope of the book, interested readers, can refer to the reference [10].

6.2 Binary Orthogonal Regression Analysis

The following uses an example to illustrate the basic method of Binary Orthogonal Regression Analysis,

【Example 6.2.1】 In the development of acid chloride ballast cement, obtained by orthogonal design its 28-day compressive strength and clinker incorporation and stone content test data are shown in Table 6.2.1, now carry out binary orthogonal regression analysis.

Table 6.2.1 Test results of 28-day strength and mixture of clinker

Compressive strength y (kg/cm²)	122	140	178	159	179	220	174	217	249
Clinker volume x_1	10	10	10	15	15	15	20	20	20
Gypsum volume x_2	4	8	12	4	8	12	4	8	12

As can be seen from the data in Table 6.2.1, the larger the dosage of clinker and the higher the intensity, the larger the amount of stone block and the higher the strength are. So imagine that they have the following relationship:

$$\hat{y} = a + b_1 x_1 + b_2 x_2 \tag{6.2.1}$$

By the least square principle, it can be derived b_1, b_2 must satisfy the following system of equations

$$\left.\begin{array}{l} l_{10} = b_1 l_{11} + b_2 l_{12} \\ l_{20} = b_1 l_{21} + b_2 l_{22} \end{array}\right\} \tag{6.2.2}$$

in which

$$l_{11} = \sum_{i=1}^{n} (x_{1i} - \bar{x}_1)^2$$

$$l_{22} = \sum_{i=1}^{n} (x_{2i} - \bar{x}_2)^2$$

$$l_{12} = l_{21} = \sum_{i=1}^{n} (x_{1i} - \bar{x}_1)(x_{2i} - \bar{x}_2)$$

$$l_{10} = \sum_{i=1}^{n} (x_{1i} - \bar{x}_1)(y_i - \bar{y})$$

$$l_{20} = \sum_{i=1}^{n} (x_{2i} - \bar{x}_2)(y_i - \bar{y})$$

$$x_1 = \frac{1}{n}\sum_{i=1}^{n} x_{1i}$$

$$x_2 = \frac{1}{n}\sum_{i=1}^{n} x_{2i}$$

$$y = \frac{1}{n}\sum_{i=1}^{n} y_i$$

The constant term a is

$$a = y - b_1 x_1 - b_2 x_2 \tag{6.2.3}$$

Therefore, to compute a, b_1, b_2, the above statistics must first be calculated. For convenience of calculation, let

$$x'_{1i} = x_{1i} - 15,\ x'_{2i} = x_{2i} - 8,\ y'_2 = x_i - 182$$

As mentioned before, a series of data, with the subtraction of a number does not affect the calculation of the sum of squares, so the x'_{1i}, x'_{2i} and y'_i calculate $l_{ij}(i, j = 0, 1, 2)$, the result is the same as the original value. Calculated as shown in Table 6.2.2 format.

Table 6.2.2 Two element regression calculation table

No.	x'_{1i}	x'_{2i}	y'_i	$(x'_{1i})^2$	$(x'_{2i})^2$	$(y')^2$	$x'_{1i} x'_{2i}$	$x'_{1i} y'_i$	$x'_{2i} y'_i$
1	−5	−4	−60	25	16	3 600	20	300	240
2	−5	0	−42	25	0	1 764	0	210	0
3	−5	4	−4	25	16	16	−20	20	−15
4	0	−4	−23	0	16	529	0	0	92
5	0	0	−3	0	0	9	0	0	0
6	0	4	38	0	16	1 444	0	0	152
7	5	−4	−8	25	16	64	−20	−40	32
8	5	0	35	25	0	1 225	0	175	0
9	5	4	67	25	16	4 489	20	335	268
Σ	0	0	0	150	96	13 140	0	1 000	768

The format of Table 6.2.2 is similar to the format of Table 6.1.2. From the last line of this table, we can calculate the amount we want

$n = 9$

$\sum y' = 0,\ \bar{y}' = (\sum y')/n = 0,\ \bar{y} = 182$

$\sum x'_1 = 0,\ \bar{x}'_1 = (\sum x'_1)/n = 0,\ \bar{x}_1 = 15$

$\sum x'_2 = 0,\ \bar{x}_2 = (\sum x'_2)/n = 0,\ \bar{x}_2 = 8$

$l_{00} = \sum (y' - \bar{y}')^2 = \sum y'^2 - \frac{1}{n}(\sum y')^2 = 13\ 140$

$l_{11} = \sum (x'_1 - \bar{x}'_1)^2 = \sum x'^2_1 - \frac{1}{n}(\sum x'_1)^2 = 150$

$$l_{22} = \sum (x'_2 - \bar{x}'_2)^2 = \sum x'^2_2 - \frac{1}{n}(\sum x'_2)2 = 96$$

$$l_{12} = l_{21} = \sum (x'_1 - \bar{x}'_1)(x'_2 - \bar{x}'_2) = \sum x'_1 x'_2 - \frac{1}{n}(\sum x'_1)(\sum x'_2) = 0$$

$$l_{10} = \sum (x'_1 - \bar{x}'_1)(y' - y') = \sum x'_1 y - \frac{1}{n}(\sum x'_1)(\sum y) = 1\,000$$

$$l_{20} = \sum (x'_2 - x'_2)(y' - \bar{y}')^2 = \sum x'_2 y - \frac{1}{n}(\sum x'_2)(\sum y) = 768$$

Thus, the regression coefficient should be satisfied for the system of equations

$$\left. \begin{array}{l} 150 b_1 + 0 \times b_2 = 1\,000 \\ 0 \times b_1 + 96 \times b_2 = 768 \end{array} \right\}$$

Solution equations, you can get

$$b_1 = \frac{1\,000}{150} = 6.66, \quad b_2 = \frac{768}{96} = 8$$

Constant $a = \bar{y} - b_1 x_1 - b_2 x = 182 - 6.66 \times 15 - 8 \times 8 = 18.1$

The required equation is

$$y = 18.1 + 6.66 x_1 + 8 x_2$$

Does this regression equation reflect the objective law of the effect? Similar to unary, it can also be measured by the amount R of the total correlation coefficient, whose meaning is exactly the same as that of the unary correlation coefficient, except that no negative value is taken $1 \geqslant R \geqslant 0$. R can be written as

$$R = \sqrt{\frac{S_N}{l_{00}}} \tag{6.2.4}$$

S_H is the sum of squares regression, which shows the change of dependent variable y due to the change of independent variables x_1 and x_2. S_N back to the commonly used formula

$$S_N = b_1 l_{10} + b_2 l_{20} \tag{6.2.5}$$

Bring in the data, get:

$$S_N = 6.66 \times 1\,000 + 8 \times 768 = 12\,810$$

so

$$R = \sqrt{\frac{12\,810}{13\,140}} = 0.987$$

R value is very close to 1, so the resulting regression equation is more ideal.

The accuracy of the regression equation is measured by the residual mean square error, S, as follows count:

$$R = \sqrt{\frac{S_{yu}}{n - m - 1}} \tag{6.2.6}$$

in which

$$S_{yu} = l_{00} - l_N \tag{6.2.7}$$

m represents the number of independent variables, where $m = 2$.

Put the value into the formula above

$S_{yu} = 13\,140 - 12\,180 = 330$

$S = \sqrt{\dfrac{330}{9-2-1}} = 7.4 \text{ kg/cm}^2$

The analysis of variance of the regression equation is shown in Table 6.2.3. Visible, regression equation is particularly remarkable.

Table 6.2.3 Variance analysis table

Variation source	Sum of squares	Freedom	Mean square	F value	Critical value
Regression	12 810	2	6 405	116.5	$F_{0.01}(2,6) = 10.92$
Residual difference	330	6	55		
SUM	12 140	8			

In the binary regression analysis, the impact of the two factors on the intensity is often not equal; there must be a major contradiction. So which one is the main contradiction? To solve this problem, we must make a significant test of the factors. The test is to compare their squared partial sum (degree of freedom 1) with the mean square of the error term. The partial regression squared sum p_i, belonging to a particular independent variable x_i, refers to the value that, in the regression equation, removes the self-deflection and reduces the sum of squared returns. For binary regression, the partial regression sum is:

$$p_1 = b_1^2 (l_{11} - \dfrac{l_{12}}{l_{22}}) \tag{6.2.8}$$

$$p_2 = b_2^2 (l_{22} - \dfrac{l_{12}}{l_{11}}) \tag{6.2.9}$$

Bring the value into, get:

$p_1 = 6.66^2 \times (150 - \dfrac{0}{96}) = 6\,666$

$p_2 = 8^2 \times (96 - \dfrac{0}{150}) = 6\,144$

The primary and secondary factors of the factors are $p_1 > p_2$, that is, the amount of clinker is greater than the amount of gypsum. As a result, calculate:

$F_1 = \dfrac{p_1}{S} = \dfrac{6\,666}{55} = 121.2^{**}$

$F_2 = \dfrac{p_2}{S} = \dfrac{6\,144}{55} = 111.7^{**}$

because $F_{0.01}(1,6) \doteq 13.75$

Therefore, both factors are particularly significant. In the experimental range, when given a clinker Jane caution dung, the root regression equation can be obtained for predicting the compressive strength of ashes, in 95% of the cases, the error is not Will exceed $2S = 14.8 \text{ kg/cm}^2$.

The reader may have found that in this case, $l_{12} = l_{21} = 0$ the regression coefficients are easily solved. This is by no means a coincidence but a regression analysis of the data derived from the orthogonal design. All data that is obtained by orthogonal design has $l_{12} = l_{21} = 0$, so that this property brings great benefits to us when there are more independent variables. The following section on this will be further explained.

6.3 Multivariate Orthogonal Regression Analysis

6.3.1 The Basic Formula

Multivariate Orthogonal Regression Similar to Binary Orthogonal Regression, most formulas can be set use.

Consider self-change: $x_1, x_2, \cdots, x_m$, dependent variable y, a total of n trials, x_{ik} represents the value of the argument x_i at the kth test, and y_k represents the result of the kth time of the dependent variable y.

Let

$$\left. \begin{aligned} l_{ij} &= \sum_{k=1}^{n}(x_{ik}-x_i)(x_{jk}-x_j), \quad i,j=1,2,\cdots,m \\ l_{i0} &= \sum_{k=1}^{n}(x_{ik}-x_i)(y_k-y), \quad i,j=1,2,\cdots,n \\ l_{00} &= \sum_{k=1}^{n}(y_k-y) \end{aligned} \right\} \quad (6.3.1)$$

in which

$$\bar{x}_i = \frac{1}{n}\sum_{k=1}^{n}x_{ik}^2, \quad i=1,2,\cdots,m$$

$$\bar{y} = \frac{1}{n}\sum_{k=1}^{n}y_k$$

If there is a linear relationship between y and x_i, the regression equation is

$$y = a + b_1 x_1 + b_2 x_2 + \cdots + b_m x_m \quad (6.3.2)$$

The regression coefficients $b_1, b_2, \cdots, b_m$ are given by the following equations:

$$\left. \begin{aligned} l_{11}b_1 + l_{12}b_2 + \cdots + l_{1m}b_m &= l_{10} \\ l_{21}b_1 + l_{22}b_2 + \cdots + l_{2m}b_m &= l_{20} \\ &\vdots \\ l_{m1}b_1 + l_{m2}b_2 + \cdots + l_{mm}b_m &= l_{m0} \end{aligned} \right\} \quad (6.3.3)$$

The constant

$$a = \bar{y} - \sum_{i=1}^{m} b_i \bar{x}_i \quad (6.3.4)$$

The sum of squares and residual squares is:

$$S_N = \sum_{i=1}^{m} l_{1i} b_4 \quad (6.3.5)$$

$$S_{yu} = l_{00} - S_N \quad (6.3.6)$$

Let the correlation coefficient is:

$$R = \sqrt{\frac{S_N}{l_{00}}} \quad (6.3.7)$$

Chapter 6 Regression Analysis of Orthogonal Design

The remaining mean square error is:

$$R = \sqrt{\frac{S_{yu}}{N - M - 1}} \tag{6.3.8}$$

As with binary regression, we use R to measure the effect of regression and S to estimate the regression process accuracy. The concept of seeking the inverse square and the inverse matrix is as follows:

Find the matrix

$$L = (l_{ij}) = \begin{pmatrix} l_{11} & l_{12} & \cdots & l_{1m} \\ l_{21} & l_{22} & \cdots & l_{2m} \\ \vdots & \vdots & & \vdots \\ l_{m1} & l_{m2} & \cdots & l_{mm} \end{pmatrix}$$

Inverse matrix

$$C = (C_{ij}) = \begin{pmatrix} C_{11} & C_{12} & \cdots & C_{1m} \\ C_{21} & C_{22} & \cdots & C_{2m} \\ \vdots & \vdots & & \vdots \\ C_{m1} & C_{m2} & \cdots & C_{mm} \end{pmatrix}$$

Partial return square sum is

$$p_i = \frac{b_i^2}{C_{ii}} \tag{6.3.9}$$

In binary, we find

$$C_{ii} = \frac{1}{(l_{ii} - \frac{l_{12}^2}{l_{ij}})}, \quad i \neq j;\ i, j = 1, 2 \tag{6.3.10}$$

From this, it is possible to calculate similarly

$$F = \frac{p_i}{s_e} \tag{6.3.11}$$

Use it to test the significance of the factors.

When the number of self-adapting fi is large, the computation of multiple regressions is quite astonishing and must be done electronically. Then to prove that,

$$l_{ij} = 0,\ i \neq j,\ i, j = 1, 2, \cdots, m$$

Thus, the system of equations 6.3.3 becomes m unary equations, which greatly simplifies the calculation. Can be seen, the test by orthogonal design, the data analysis is very favorable. Here's an example of a ternary regression based on orthogonal design.

6.3.2 Calculate the Example

【Example 6.3.1】 The slump test results obtained by orthogonal design at different gray water ratios, water consumption, and different dosages of the water reducer FDN are shown in Table 6.3.1 Slump of gray water ratio x_1, water consumption x_2 and FDN content x_3 for three yuan regression analysis.

Orthogonal Design in Concrete Application

Table 6.3.1 Test results of slump and water cement ratio, water volume, FDN volume

water cement ratio x_1	2.5	3.0	3.5	2.5	3.0	3.5	2.5	3.0	3.5
Water volume x_2 (kg/m³)	155	155	155	150	150	150	145	145	145
FDN volume x_3 (%)	1.2	0.8	1.0	1.0	1.2	0.8	0.8	1.0	1.2
Slump y (cm)	20.1	10.8	8.9	17.3	14.1	2.8	1.0	3.8	6.3

(Experimental results of the Ministry of Communications Science Research Institute and the 00069 force)

The relation of each factor to the collapse is like Figure 6.3.1.

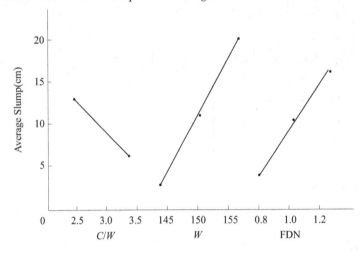

Figure 6.3.1 *The relationship between collapse degree and C/W, W and FDN*

The abscissa of the graph is the three levels of factors, and the vertical vibration is the average slump for each level. This shows that all can be considered as a straight line. So, we assume with a linear regression equation

$$y = a + b_1 x_1 + b_2 x_2 + b_3 x_3$$

For calculation convenience, gray-water ratio, water consumption and FDN miserable respectively, minus their average, because the level of each factor is equally spaced, the average $\bar{x}_i$ of x_i, is the middle of that level, that is

$$x'_{1i} = x_{1i} - 3, \quad x'_{2i} = x_{2i} - 150, \quad x'_{3i} = x_{3i} - 1.0$$

The calculation is in the form of Table 6.3.2. From the last line of the table, we calculate the statistics we need:

$n = 9$

$\sum y = 85.5, \quad \sum x'_1 = 0, \quad \sum x'_2 = 0, \quad \sum x'_3 = 0,$

$\bar{y} = 9.5, \quad \bar{x}_1 = 3.0, \quad \bar{x}_2 = 150, \quad \bar{x}_3 = 1.0$

$l_{00} = \sum y^2 - \dfrac{1}{n}\left(\sum y\right)^2 = 1\,175.21 - \dfrac{1}{9} \times 85.5^2 = 362.96$

$l_{11} = \sum x'^2_1 - \dfrac{1}{n}\left(\sum x'_1\right)^2 = 1.5 - \dfrac{1}{9} \times 0 = 1.5$

Chapter 6 Regression Analysis of Orthogonal Design

Table 6.3.2 Three element regression calculation table

No.	x'_1	x'_2	x'_3	y	x'^2_1	x'^2_2	x'^2_3	No.	y^2	$x'_1 x'_2$	$x'_1 x'_3$	$x'_2 x'_3$	$x'_1 y$	$x'_2 y$	$x'_3 y$
1	−1.5	5	0.2	20.1	0.25	25	0.04	1	404.01	−2.5	−0.1	1	−10.05	100.5	4.02
2	0	5	−0.2	10.8	0	25	0.04	2	116.64	0	0	−1	0	54	−2.16
3	0.5	5	0	9.8	0.25	25	0	3	96.04	2.5	0	0	4.9	49	0
4	−0.5	0	0	17.3	0.25	0	0	4	299.29	0	0	0	−8.65	0	0
5	0	0	0.2	14.1	0	0	0.04	5	198.81	0	0	0	0	0	2.82
6	0.5	0	−0.2	2.3	0.25	0	0.04	6	5.29	0	−0.1	0	1.15	0	−0.46
7	−0.5	−5	−0.2	1.0	0.25	25	0.04	7	1.00	2.5	0.1	1	−0.5	−5	−0.20
8	0	−5	0	3.8	0	25	0	8	14.44	0	0	0	0	−19	0
9	0.5	−5	0.2	6.3	0.25	25	0.04	9	39.69	−2.5	0.1	−1	3.4	−34	1.36
∑	0	0	0	85.5	1.50	150	0.24	∑	1175.21	0	0	0	−9.75	145.5	5.38

$$l_{22} = \sum (x'_2 - \bar{x}'_2)^2 = \sum x'^2_2 - \frac{1}{n}(\sum x'_2)^2 = 150 - \frac{1}{9} \times 0 = 150$$

$$l_{33} = \sum (x'_3 - \bar{x}'_3)^2 = \sum x'^2_3 - \frac{1}{n}(\sum x'_3)^2 = 0.24 - \frac{1}{9} \times 0 = 0.24$$

$$l_{12} = l_{21} = \sum x'_1 x'_2 - \frac{1}{n}(\sum x'_1)(\sum x'_2) = 0$$

$$l_{13} = l_{31} = \sum x'_1 x'_3 - \frac{1}{n}(\sum x'_1)(\sum x'_3) = 0$$

$$l_{23} = l_{32} = \sum x'_2 x'_3 - \frac{1}{n}(\sum x'_2)(\sum x'_3) = 0$$

$$l_{10} = \sum x'_1 y - \frac{1}{n}(\sum x'_1)(\sum y) = -9.75 - \frac{1}{9} \times 0 = -9.75$$

$$l_{20} = \sum x'_2 y - \frac{1}{n}(\sum x'_2)(\sum y) = 145.5 - \frac{1}{9} \times 0 = 145.5$$

$$l_{30} = \sum x'_3 y - \frac{1}{n}(\sum x'_3)(\sum y) = 5.38 - \frac{1}{9} \times 0 = 5.38$$

Thus, the regression coefficient should satisfy the system of equations

$$\left. \begin{aligned} 1.5 b_1 + 0 \times b_2 + 0 \times b_3 &= -9.75 \\ 0 \times b_1 + 150 \times b_2 + 0 \times b_3 &= 145.5 \\ 0 \times b_1 + 0 \times b_2 + 0.24 \times b_3 &= 5.38 \end{aligned} \right\}$$

Calculate equations,

$$b_1 = \frac{-9.75}{1.5} = -6.5$$

$$b_2 = \frac{145.5}{150} = 0.97$$

$$b_3 = \frac{5.38}{0.24} = 22.42$$

The constant

$$\begin{aligned} a &= \bar{y} - b_1 \bar{x}_1 - b_2 \bar{x}_2 - b_3 \bar{x}_3 \\ &= 9.5 + 6.5 \times 3 - 0.97 \times 150 - 22.42 \times 1.0 \\ &= -138.92 \end{aligned}$$

The equation we want is

$$\hat{y} = -138.92 - 6.5 x_1 + 0.97 x_2 + 22.42 x_3$$

Regression sum of squares

$$\begin{aligned} S_N &= b_1 l_{10} + b_2 l_{20} + b_3 l_{30} \\ &= -6.5 \times (-9.75) + 0.97 \times 145.5 + 22.42 \times 5.38 \\ &= 325.12 \end{aligned}$$

Full correlation coefficient

$$R = \sqrt{\frac{S_N}{l_{00}}} = \sqrt{\frac{325.12}{362.96}} = 0.946$$

Remaining square sum

Chapter 6 Regression Analysis of Orthogonal Design

$S = l_{00} - S_N = 362.96 - 325.12 = 37.84$

The remaining mean square error

$$S = \sqrt{\frac{S_{yu}}{n-m-1}} = \sqrt{\frac{37.84}{9-3-1}} = 2.75$$

The diagonal matrix of normal equation coefficients is

$$L = (l_{ij}) = \begin{pmatrix} l_{11} & 0 & 0 \\ 0 & l_{22} & 0 \\ 0 & 0 & l_{33} \end{pmatrix} = \begin{pmatrix} 1.5 & 0 & 0 \\ 0 & 1.5 & 0 \\ 0 & 0 & 1.5 \end{pmatrix}$$

The inverse matrix is

$$C = \begin{pmatrix} 1/l_{11} & 0 & 0 \\ 0 & 1/l_{22} & 0 \\ 0 & 0 & 1/l_{33} \end{pmatrix} = \begin{pmatrix} 1/1.5 & 0 & 0 \\ 0 & 1/150 & 0 \\ 0 & 0 & 1/0.24 \end{pmatrix}$$

Partial return to the sum of squares

$$p_1 = \frac{b_1^2}{c_{11}} = \frac{-6.5^2}{0.6666} = 63.375$$

$$p_2 = \frac{b_2^2}{c_{22}} = \frac{0.97^2}{0.0067} = 141.135$$

$$p_3 = \frac{b_3^2}{c_{33}} = \frac{22.42^2}{4.1667} = 120.638$$

An analysis of variance is given in Table 6.3.3.

From variance analysis we can see:

1. Various factors affect the slump order of primary and secondary $x_1 \to x_2 \to x_3$, that is, water consumption - FDN content-gray water ratio.
2. Since the influence of water consumption on slump is a particularly significant factor, and the FDN and gray water ratios are significant factors, the regression of slump on three factors is particularly significant.

Table 6.3.3 Variance analysis table

Variation source	Sum of squares	Freedom	Mean square	F value	Critical value
Regression	325.12	3	108.37	14.3	$F_{0.01}(3,5) = 12.06$
x_1 (C/W)	53.375	1	63.37	8.37**	$F_{0.01}(1,5) = 16.26$
x_2 (W)	141.135	1	141.135	18.64*	$F_{0.05}(1,5) = 6.61$
x_3 (FDN)	120.638	1	120.638	15.94**	
Surplus	37.84	5	7.57		
SUM	362.98	8			

3. Due to the small number of tests, the test error is still large, $s = 2.75$ cm. This example is not repeated test results, only to illustrate the method of multiple orthogonal regression analysis.

Chapter 7
Data Structure and Effect Estimation of Orthogonal Design

The data structure is the basis of orthogonal design data processing, and the data structure is clarified. Not only the mathematical principle of orthogonal design visual analysis can be understood in theory, but also the effect estimation in orthogonal design can be solved, that is, the optimal theoretical value The estimation problem. This chapter will briefly introduce the basic concepts of data structure, the data structure of orthogonal design and the theoretical value of the optimal process. Readers who want to know more about the above issues can refer to [1,11].

7.1 Basic Concepts of Data Structure

As we all know, concrete test results (data) are always different (fluctuating), even under the same conditions, repeated tests, the data will not be the same. This fluctuation of the data is caused by the following reasons: ① the difference of the mixture ratio; ② the difference of the process conditions; ③ the operation of the personnel and the measurement of disease and so on.

As mentioned earlier, although there are many factors that affect the testing of culinary diseases, they have an unequal impact on assessment indicators. There are major and minor diseases in Lang. The purpose of orthogonal design is to examine some of the more important factors that led to the capture of some of the affected areas by Qin media so that these controlled factors are fixed at the level we hope. Theoretically, their effect on the test result x is fixed, and we denote these effects in m. For minor factors, there is usually no control (and sometimes difficult to control). Their impact on the test results can be summed up as one, called "error term", we use ε said. Then any test result (data) x can be decomposed into two parts:

$$x = m + \varepsilon \qquad (7.1.1)$$

Where m said the various controlled factors on the impact of the sum of the index, that is, in a certain ratio, under the conditions of a study index should have the theoretical value. ε represents the sum of the effects of many uncontrolled factors (or stochastic factors) on the data x from the ingredients to the entire test, and we call ε a random error.

From statistical mathematics we can see that m and ε are two types of different nature, the former is common, the latter is a random variable. Because under certain conditions, the value of ε is an indefinite. If we add a random variable ε to a constant m, the resulting result is also a random

Chapter 7 Data Structure and Effect Estimation of Orthogonal Design

variable. Obviously, the randomness of x is caused by the random error ε. Equation 7.1.1 is called the data structure of x. To illustrate its significance, the single-factor repeated trial data in Example 3.1.1 was decomposed.

The results of the repeated test will be copied at different dosages in Example 3.1.1 and are listed in Table 7.1.1.

Table 7.1.1

Repeated test times	P_1	P_2	P_3	P_4
1	$x_{11} = 185$	$x_{21} = 173$	$x_{31} = 154$	$x_{41} = 145$
2	$x_{12} = 164$	$x_{22} = 148$	$x_{32} = 140$	$x_{42} = 140$
3	$x_{13} = 192$	$x_{23} = 180$	$x_{33} = 169$	$x_{43} = 158$
4	$x_{14} = 200$	$x_{24} = 174$	$x_{34} = 170$	$x_{44} = 168$
Average strength (kg/cm^2)	185.25	168.75	158.25	152.75
Total mean value (kg/cm^2)	166.25			

The data x_{ij} in the table shows the intensity of the j-th test at the i-th level. Obviously, it is a random variable, which can be decomposed into

$$x_{ij} = m_i + \varepsilon_{ij} \quad (i = 1, 2, \cdots, p, j = 1, 2, \cdots, p)$$

Where m_i reflects the i-th dosage should have the intensity value, ε_{ij} said that due to various reasons, random error, p said the number of levels, r said the number of repeated tests at each level. This example $p = 4$, $r = 4$.

For further discussion, we need to introduce the concepts of "average" and "effects", which we claim

$$\mu = \frac{1}{p} \sum_{i=1}^{p} m_i \quad (i = 1, 2, \cdots, p) \tag{7.1.2}$$

As a general average, this factor can be interpreted as a "medium" test result.

$$a_i = m_i - \mu \quad (i = 1, 2, \cdots, p) \tag{7.1.3}$$

The i-th level of effect is called the factor. It describes how much or how bad a test result for this factor is at the i-th level. So we wrote the data structure

$$x_{ij} = \mu + a_i + \varepsilon_{ij} \quad (i = 1, 2, \cdots, p; j = 1, 2, \cdots, r) \tag{7.1.4}$$

Obviously, a_i should satisfy the following formula

$$\sum_{i=1}^{p} a_i = \sum_{i=1}^{p} (m_i - \mu) = \sum_{i=1}^{p} m_i - p\mu = 0 \tag{7.1.5}$$

For this example,

$\mu = 166.25$

$a_1 = 185.25 - 166.25 = 19$

$a_2 = 168.75 - 166.25 = 2.5$

$a_3 = 158.25 - 166.25 = -8$

$a_4 = 152.75 - 166.25 = -13.5$

we can see, $a_1 + a_2 + a_3 + a_4 = 0$

$\varepsilon_{11} = x_{11} - \mu - a_1 = 185 - 166.25 - 19 = -0.25$

$\varepsilon_{12} = x_{12} - \mu - a_1 = 164 - 166.25 - 19 = -21.25$

$\varepsilon_{13} = x_{13} - \mu - a_1 = 192 - 166.25 - 19 = 6.25$

$\varepsilon_{14} = x_{14} - \mu - a_1 = 200 - 166.25 - 19 = 14.75$

In this way, x_{ij} can be decomposed into three numbers, that is

$x_{11} = 166.25 + 19 - 0.25$

$x_{12} = 166.25 + 19 - 21.25$

$x_{13} = 166.25 + 19 + 6.25$

$x_{14} = 166.25 + 19 + 14.75$

Other data pages can be similarly decomposed to obtain Table 7.1.2.

Table 7.1.2

Repeated test times	p_1	p_2	p_3	p_4
1	185 = 166.25 +19 −0.25	173 = 166.25 +2.5 +4.25	154 = −8 −4.25	145 = 166.25 −13.5 −7.75
2	164 = 166.25 +19 −6.75	148 = 166.25 +2.5 −20.75	140 = 166.25 −8 −18.25	140 = 166.25 −13.5 −12.75
3	192 = 166.25 +19 +6.75	180 = 166.25 +2.5 +11.25	169 = 166.25 −8 +10.75	158 = 166.25 −13.5 +5.25
4	200 = 166.25 +19 +14.75	174 = 166.25 +2.5 +5.25	170 = 166.25 −8 +11.75	168 = 166.25 −13.5 +15.25

Through the decomposition of each data, we can see the size of the effect of A (the impact of dosage) and the size of the error.

7.2 Data Structure Orthogonal Design

In the orthogonal design, the test conditions of different numbers of test results are different, the data structure of the resulting results are different. For example, in Example 5.1.1, two factors A and B (see Table 7.2.1) are arranged on the orthogonal table $L_4(2^3)$, and the test results are x_1, x_2, x_3, x_4.

Chapter 7 Data Structure and Effect Estimation of Orthogonal Design

Table 7.2.1

test times	A	B		Experiment result
	1	2	3	
1	1	1	1	x_1
2	1	2	2	x_2
3	2	1	2	x_3
4	2	2	1	x_4

In each number of tests, because of the different levels of the factors, the sum of the impact of these factors on the indicators is not the same. I use m_{11}, m_{12}, m_{21} and m_{22} to show the correlation between these factors and the indicators x_1, x_2, x_3, x_4 the sum of the impact. The random error in each test followed by $\varepsilon_1, \varepsilon_2, \varepsilon_3, \varepsilon_4$, so

$$\left.\begin{array}{l} x_1 = m_{11} + \varepsilon_1 \\ x_2 = m_{12} + \varepsilon_2 \\ x_3 = m_{13} + \varepsilon_3 \\ x_4 = m_{14} + \varepsilon_4 \end{array}\right\} \quad (7.2.1)$$

This set of data structures is called the mathematical model for arranging experiments on $L_4(2^3)$. Which represents the theoretical value of the test results under the condition that factor A takes i level and factor B takes j level.

Similar to the single factor, we can also decompose mij further

$$\mu = \frac{1}{pq} \sum_{i=1}^{p} \sum_{j=1}^{q} m_{ij} \quad (7.2.2)$$

That the various factors to take the theoretical level of test results when the median, still said the general level. Where p, q, respectively, the level of the number of factors A, B (in this case $p = 2$, $q = 2$); and

$$\mu_i = \frac{1}{q} \sum_{j=1}^{q} m_{ij} \quad (i = 1, 2) \quad (7.2.3)$$

$$\mu = \frac{1}{pq} \sum_{i=1}^{p} m_{ij} \quad (j = 1, 2) \quad (7.2.4)$$

Respectively, the theoretical value of the test results when factor A takes i level and the theoretical value of test result when factor B takes i level,

$$a_i = \mu_i - \mu \quad (i = 1, 2) \quad (7.2.5)$$
$$b_j = \mu_j - \mu \quad (j = 1, 2) \quad (7.2.6)$$

Respectively, when the factor A take i level effect and factor B take j level effect. Obviously, there is a relationship between a_i and b_j, respectively,

$$\sum_{i=1}^{p} a_i = \sum_{i=1}^{p} (\mu_i - \mu) = \sum_{i=1}^{p} \mu_i - p\mu = 0 \quad (7.2.7)$$

$$\sum_{j=1}^{q} b_j = \sum_{j=1}^{q} (\mu_j - \mu) = \sum_{j=1}^{q} \mu_j - q\mu = 0 \quad (7.2.8)$$

Orthogonal Design in Concrete Application

The following two cases were written $L_4(2^3)$ test results of the data structure.

a. $m_{ij} = \mu + a_i + b_j$ That is, second case in case 5.1.1 (No interaction). This situation illustrates the impact of factors A, B on the data by their horizontal effect superimposed, the data structure can be written as:

$$\left.\begin{aligned}x_1 &= \mu + a_1 + b_1 + \varepsilon_1 \\ x_2 &= \mu + a_1 + b_2 + \varepsilon_2 \\ x_3 &= \mu + a_2 + b_1 + \varepsilon_3 \\ x_4 &= \mu + a_2 + b_2 + \varepsilon_4\end{aligned}\right\} \quad (7.2.9)$$

With the data structure of Equation 7.2.9, you can gain a deeper understanding of why intuitive analysis can be used to analyze the range. From 7.2.9 can be introduced:

For the first column of the arrangement,

$$K_1^A - K_2^A = (x_1 + x_2) - (x_3 + x_4) = 2(a_1 + a_2) + (\varepsilon_1 + \varepsilon_3) + (\varepsilon_2 + \varepsilon_4)$$

Only reflects the difference between the two levels of factor A (does not contain any component of factor B) and the difference caused by random error. Similarly, for the second column occupied by Factor B, there is

$$K_1^B - K_2^B = (x_1 + x_3) - (x_2 + x_4) = 2(b_1 - b_2) + (\varepsilon_1 + \varepsilon_3) - (\varepsilon_2 + \varepsilon_4)$$

It also reflects only the difference between the two levels of B (excluding any component of factor A) and the difference caused by random errors. Thus, visual analysis can use $K_1 - K_2$ to reflect the difference between the two factors A or B levels. At the same time see, in the mixed part of the impact of the error, so the accuracy is poor. From here you can also further recognize the need for analysis of variance.

For empty columns (column 3), there is

$$K_1^N - K_2^N = (x_1 + x_4) - (x_2 + x_3) = (\varepsilon_1 + \varepsilon_4) - (\varepsilon_2 + \varepsilon_3)$$

It reflects the difference caused by the error, without any impact of factors A or B. The effect of partial error between $K_1 - K_2$. Therefore, the margins of empty columns reflect the experimental error.

b. $m_{ij} \neq \mu + a_i + b_j$ Case 5.1.1 first case (with interaction). This situation shows that the impact of factors A, B data is not simply the superposition of their respective effects, but also must consider the interaction between the two factors A, B on the data, we call

$$(ab)_{ij} = m_{ij} - \mu - a_i - b_j \quad (i = 1, 2; j = 1, 2)$$

The above formula is called the interaction between factor A and B, and take i level in factor A, and factor B takes the effect of j as the level of interaction. Obviously, there is a relation to $(ab)_{ij}$:

$$\sum_{i=1}^{p}(ab)_{ij} = \sum_{i=1}^{p}(m_{ij} - \mu - a_i - b_j) = p\mu_j - p\mu - \sum_{i=1}^{p}a_i - pb_j \quad (7.2.10)$$
$$= p(\mu_j - \mu) - pb_j = pb_j - pb_j = 0$$

$$\sum_{j=1}^{q}(ab)_{ij} = \sum_{j=1}^{q}(m_{ij} - \mu - a_i - b_j) = q\mu_j - q\mu - \sum_{i=1}^{p}a_i - qb_j \quad (7.2.11)$$
$$= q(\mu_j - \mu) - qb_j = qb_j - qb_j = 0$$

Data structure can be written as:

Chapter 7 Data Structure and Effect Estimation of Orthogonal Design

$$\left.\begin{aligned}x_1 &= \mu + a_1 + b_1 + (ab)_{11} + \varepsilon_1 \\ x_2 &= \mu + a_2 + b_2 + (ab)_{12} + \varepsilon_2 \\ x_3 &= \mu + a_3 + b_3 + (ab)_{13} + \varepsilon_3 \\ x_4 &= \mu + a_4 + b_4 + (ab)_{14} + \varepsilon_4 \end{aligned}\right\} \quad (7.2.12)$$

According to Formula. 7.2.12, it is easy to introduce visual analysis with interaction and also use range analysis. At this moment, the range of empty columns reflects the interaction between A and B: $A \times B$.

7.3 Orthogonal Designs

The effect estimation is a prediction problem to solve the theoretical value of the optimal process (or formula), which is a concrete application of the data structure. In concrete tests, sometimes the indicators are not evaluated as high as possible (or the lower the better), but hope that the test indicators can meet the design requirements or reach a specific value, then we must answer the choice of the most What is the average theoretical value m of a good process (or recipe) For example, the third experiment in Example 5.3.1 was performed under the combination of $A_1 B_2 C_1$, Its 7-day strength is 273 kg/cm^2. Since the total randomness of an experiment, we ask: What is the average intensity of a large number of repeated tests? Now we can answer this question using the orthogonal design data structure.

The test plan and test results in Table 5.3.3 are copied and listed in Table 7.3.1. From the analysis know, $A \times C$, $B \times C$ has no effect on the strength. Therefore, only the factors A, B, C and the interaction $A \times B$ are examined on the $L_8(2^7)$ orthogonal table.

According to Table 7.3.1, write out its data structure:

$$\left.\begin{aligned}x_1 &= m_1 + \varepsilon_1 = \mu + a_1 + b_1 + (ab)_{11} + C_1 + \varepsilon_1 \\ x_2 &= m_2 + \varepsilon_2 = \mu + a_1 + b_1 + (ab)_{11} + C_2 + \varepsilon_2 \\ x_3 &= m_3 + \varepsilon_3 = \mu + a_1 + b_2 + (ab)_{12} + C_1 + \varepsilon_3 \\ x_4 &= m_4 + \varepsilon_4 = \mu + a_1 + b_2 + (ab)_{12} + C_2 + \varepsilon_4 \\ x_5 &= m_5 + \varepsilon_5 = \mu + a_2 + b_1 + (ab)_{21} + C_1 + \varepsilon_5 \\ x_6 &= m_6 + \varepsilon_6 = \mu + a_2 + b_1 + (ab)_{21} + C_2 + \varepsilon_6 \\ x_7 &= m_7 + \varepsilon_7 = \mu + a_2 + b_2 + (ab)_{22} + C_1 + \varepsilon_7 \\ x_8 &= m_8 + \varepsilon_8 = \mu + a_2 + b_2 + (ab)_{22} + C_2 + \varepsilon_8 \end{aligned}\right\} \quad (7.3.1)$$

Table 7.3.1

No.	A	B	A×B	C				Experiment Result x_i
	1	2	3	4	5	6	7	
1	1	1	1	1	1	1	1	$x_1 = 169$
2	1	1	1	2	2	2	2	$x_2 = 178$
3	1	2	2	1	1	2	2	$x_3 = 273$

Continued Table 7.3.1

No.	A	B	A×B	C				Experiment Result x_i
	1	2	3	4	5	6	7	
4	1	2	2	2	2	1	1	$x_4 = 272$
5	2	1	2	1	2	1	2	$x_5 = 146$
6	2	1	2	2	1	2	1	$x_6 = 169$
7	2	2	1	1	2	2	1	$x_7 = 194$
8	2	2	1	2	1	1	2	$x_8 = 215$

From formula 7.3.1 we see that although the m_i for each experiment is not unambiguous, and they are both the general average μ and the linear sum of the various factors and the interaction effects, only the general average and the corresponding estimates should be obtained, The value of the estimate can be solved.

Let the estimated value of m_i be the $\hat{m}_i$ we hope to obtain, as close to the measured value x_i as possible. In other words, we must minimize the residual $(x_i - \hat{m}_i)$ as much as possible. For this purpose, according to the least squares farther away, the sum of squared residuals is required

$$Q = \sum_{i=1}^{n} (x_i - m_i)^2 \quad (n = 8) \tag{7.3.2}$$

From formula 7.3.1, to determine m_i, is to set μ, a_i, b_j, c_k, $(ab)_{ij}$ valuation $\hat{\mu}$, $\hat{a}_i$, $\hat{b}_j$, $\hat{c}_k$, $(\hat{ab})_{ij}$. As mentioned above, the relationship between the parameters

$$\sum_{i=1}^{2} a_i = 0, \quad \sum_{j=1}^{2} b_j = 0, \quad \sum_{k=1}^{2} c_k = 0$$

$$\sum_{i=1}^{2} (ab)_{ij} = 0 \, (j = 1,2), \quad \sum_{j=1}^{2} (ab)_{ij} = 0 \quad (i = 1,2)$$

Their valuations also satisfy these relationships, ie

$$\sum_{i=1}^{2} \hat{a}_i = 0, \quad \sum_{j=1}^{2} \hat{b}_j = 0, \quad \sum_{k=1}^{2} \hat{c}_k = 0$$

$$\sum_{i=1}^{2} (\hat{ab})_{ij} = 0 \quad (j = 1,2), \quad \sum_{j=1}^{2} (\hat{ab})_{ij} = 0 \quad (i = 1,2)$$

To minimize Q, substitute Formula (7.3.1) in Formula (7.3.2), derive partial derivatives of $\hat{\mu}$, $\hat{a}_i$, $\hat{b}_j$, $\hat{c}_k$, $(\hat{ab})_{ij}$, respectively, and make them zero. Valuation of parameters:

From $\dfrac{\partial Q}{\partial \hat{\mu}} = 0$ we can get

$$\hat{\mu} = \frac{1}{8} \sum_{i=1}^{8} x_i = \bar{x}_i$$

That is, the average μ of the valuation $\hat{\mu}$ is equal to the average of all data.

From $\dfrac{\partial Q}{\partial \hat{\mu}_1} = 0$ we can get

Chapter 7 Data Structure and Effect Estimation of Orthogonal Design

$$a_1 = \frac{x_1 + x_2 + x_3 + x_4}{4} - \bar{x}$$

Which x_1, x_2, x_3, x_4 exactly A_1 level four test results. It can be seen from the above that the effect estimate $\hat{a}_1$ when the factor A takes a level is the average of the experimental data at the A_1 level minus the total average of all the experimental data.

Similarly,

$\hat{a}_i = \hat{a}_i$, average of test data at level-x

$\hat{B}_j = \hat{B}_j$ average test data-x

$\hat{c}_k$ = average of test data at c_k level-x

$(ab)_{ij}$ = average of test data at $A_i B_j - \hat{a}_i - \hat{b}_j - \bar{x}$. The average estimated optimum process is:

$m_i = \mu$ + a good estimate of the significant factor

From the significance test result of Example 5.3.1, then

$$\hat{m}_i = \hat{\mu} + \hat{a}_1 + \hat{b}_2 + (\hat{ab})_{12}$$

According to the data in Table 7.3.1, we can get:

$$\hat{\mu} = \bar{x} = \frac{1\,616}{8} = 202$$

$$\hat{a}_1 = \frac{892}{4} - 202 = 21$$

$$\hat{b}_2 = \frac{782}{4} - 202 = 36.5$$

$$(\hat{ab})_{12} = \frac{273 + 272}{2} - 21 - 36.5 - 202 = 13$$

then $\hat{m} = 202 + 21 + 36.5 + 13 = 272.5$

That a large number of repeated tests, the average strength of 272.5 kg/cm². Due to the error in the test, the test value generally fluctuates around 272.5 kg/cm².

Easy to launch, the test No. 4 test conditions although $A_1 B_2 C_2$, the average theoretical strength should be 272.5 kg/cm², and the measured intensity of 272 kg/cm² is very close to it, therefore, from the theoretical proof, $A_1 B_2 C_1$ conditions for the process is completely reasonable (conservation time shortened 1 hour, and achieve the same effect).

Appendix I

Supplementary Examples

I.1 【Example I-1】 Preferred Naphthalene Super Plasticizer DH-3 and Other Process Parameters

Currently produced naphthalene-based reducing agent - naphthalene sulfonic acid formaldehyde condensate, roughly divided into the following categories: naphthalene as raw material NNO water reducer, methyl naphthalene as a raw material for the MF water reducer and methyl naphthalene distillation A flag for the construction of a water reducer and so on. Their water reduction rate is up to 13% - 15%. As the super plasticizer still has the role of large gas, the price is more expensive and other shortcomings, in recent years; many units of water reducer modification conducted a large number of tests. For example, NF made by Tsinghua University, FDN developed by Zhanjiang Pesticide Factory, suffocation oil produced by Shanghai Jianke Institute and Shanghai May 4th Farm, DH water reducer developed by Haikou Water Conservancy Bureau and so on. In the development process, the above units have adopted the visual analysis of orthogonal design to optimize the process parameters of water reducer. The following highlights the preferred DH-3 water-reducing agent process parameters.

The purpose of the experiment is to search for naphthalene and methylnaphthalene using washing oil, as a raw material to synthesize new naphthalene-based water-reducing agent and the composition of the naphthalene sulfonic acid formic acid condensate is still present. The performance of the naphthalene sulfonic acid formic acid condensate is better than that of a similar product. Process parameters.

Assessment indicators: mobility.

I.1.1 Factors and Levels

Based on the general experience of synthesis of naphthalene super plasticizer, affect the performance of the product process the parameters have the following ten:

1. Naphthalene in raw materials at 230 - 300 Celsius degrees: 5% - 15%;

2. The use of acid amount: 1.33 – 1.55 molar ratio;

3. Suffocation temperature: 150 to 170 Celsius degrees;

4. Suffocation time: 2 hours;

5. Water for hydrolysis: 2.86 mole ratio;

6. Hydrolysis temperature: less than or equal to 100 Celsius degrees;

7. Hydrolysis time: 15 minutes,

8. Formaldehyde condensation dosage: 0.6 – 0.8 molar ratio,

9. Condensation temperature: less than or equal to 100 Celsius degrees

10. Condensation time: 2 hours.

Among the above ten parameters, four parameters are considered to have a major influence on the synthesis of DH-3: ① naphthalene content in raw oil, ② amount of sulfuric acid used for cracking, ③ suffocation temperature, ④ the amount of formaldehyde condensation. Therefore, the choice of these four parameters as a factor, each net pick three levels, factors and levels in Table I.1.1.

Table I.1.1 Factor level table

Level	Factor			
	A. Alunite admixture (%)	B. Sulphuric acid content(mL)	C. Slag content (℃)	D. Formaldehyde content(mL)
1	5	48	150 – 153	29.1
2	10	52.3	160 – 163	34.0
3	15	56	167 – 170	38.9

I.1.2 Test Program and Test Results

This example is four factors; each factor has three levels, so use $L_9(3^4)$ to arrange the test. The test plan is shown in Table I.1.2.

In order to improve the reliability of the test results, duplicate tests were performed for each test. The results are given on the right side of Table I.1.2. The method of analysis with repeated tests is the same as Example 4.2.3.

I.1.3 Analysis of Test Results
I.1.3.1 Intuitive Analysis

1) Look directly. The maximum fluidity of test No. 6 is 410 mm. The combined conditions are $A_3B_2C_1D_2$ as follows: the naphthalene content is 15%, the amount of sodium sulphate is 52.3 mL, the suffocation temperature is 150 – 153 ℃ and the amount of formaldehyde is 34 mL.

2) Calculate the sum of each element of the column, calculate the sum of the corresponding levels of fluid mobility K_1, K_2 and K_3 and its range R, the calculation results recorded in Table I.1.2 below.

Table I.1.2 $L_9(3^4)$ **Test results and calculation**

Experimental number	A	B	C	D	Experimental result (mL)		SUM	R_i
	1	2	3	4	1	2		
1	1(5)	1(48)	3(167–170)	2(34.0)	186	190	376	4
2	2(10)	1	1(150–153)	1(29.1)	194	198	392	4
3	3(15)	1	2(160–163)	3(38.9)	193	201	394	8
4	1	2(52.8)	2	1	196	196	392	0
5	2	2	3	3	195	200	396	4
6	3	3	1	2	196	205	410	0
7	1	3(56)	1	3	205	202	406	2
8	2	3	2	2	205	196	396	4
9	3	3	3	1	200	200	403	3
K_1	1 174	1 162	1 208	1 187				
K_2	1184	1 198	1 182	1 182			SUM:3 565	
K_3	1207	1 205	1 175	1 196				
R	33	43	33	14				

According to the size of R, the order of each factor is $B \rightarrow A$ and $C \rightarrow D$. That is, the amount of sulfuric acid is the main factor affecting the fluidity. The content of naphthalene and the suffocation temperature are secondary factors. The effect of formaldehyde is very small.

A possible process condition resulting from the size of the K value is $A_3 B_2 C_1 D_3$.

How to choose the best process parameters? The good condition that can be seen directly is $A_3 B_2 C_1 D_2$. A good condition to calculate is $A_3 B_2 C_1 D_3$ that it is not included in the 9 tests. In the above two combination conditions, the secondary factors A and C are the same, and the impact is very small because D selects D_2 or D_3, but the main factor B still needs to select B_2 or B_3: It still needs to be determined by analysis of variance.

I.1.3.2 Analysis of Variance

1) Estimation of Experimental Error

The difference between the two replicates of the same test number reflects the trial misalignment. The last column in Table I.1.2 gives the worst-case R_i for each trial and averages them.

$$\overline{R} = \frac{1}{9} \times (4 + 4 + \cdots + 3) = 3.22$$

In this case, $n = 2$, $l = 9$, and the appendix to the table 4 is $d(2,9) = 1.16$.

$\hat{\sigma} = \overline{R}/d(2,9) = 3.22/1.16 = 2.78$

Freedom of error degree is

$f_e = \phi(n,l) = 0.9l(n-1) = 8$

2) Test the significance of A, B, C, D

$R_A = R_C = 33, r_A = r_C = 2 \times 3 = 6$

$q_A = q_C = 33/2.78/\sqrt{6} = 4.85$

$R_B = 43, r_B = 6$

$q_B = 43/2.78/\sqrt{6} = 6.31$

$R_D = 14, r_D = 6$

$q_D = 14/2.78/\sqrt{6} = 2.06$

$q_{0.05}(3, 8) = 4.04, q_{0.01}(3, 8) = 5.63$

Variance analysis: B is particularly significant; A and C are significant; no effect of D is seen.

3) Multiple comparisons of factor B

in this case

$\overline{S}_e = \hat{\sigma}^2 = 2.78^2 = 7.70$

$f_e = 8$

$r = 2 \times 3 = 6$

$m = 3$

$\overline{K}_1 = 193.7, \overline{K}_2 = 199.7, \overline{K}_3 = 200.8$

Check q table: $q_{0.05}(m, f_e) = q_{0.05}(3, 8) = 4.04$

Calculate:

$d_T = q_{0.05}(3,8) \sqrt{\dfrac{Se}{r}} = 4.04 \sqrt{\dfrac{7.7}{6}} = 4.58$

$d_{12} = |\overline{K}_1 - \overline{K}_2| = |193.7 - 199.7| = 6$

$d_{13} = |\overline{K}_1 - \overline{K}_3| = |193.7 - 200.8| = 7.1$

$d_{23} = |\overline{K}_2 - \overline{K}_3| = |199.7 - 200.8| = 1.1$

It can be seen that the level 1 of factor B is significantly different from level 2 and level 3, while there is no significant difference between level 2 and level 3.

From the results of multiple comparisons, since there is no significant difference between B_2 and B_3, the two can be either. Finally, the optimal combination condition is $A_3 B_2 (B_3) C_1 D_0$. D_o represents any of the three levels of factor D. That is, the naphthalene content is 15%, the sulfuric acid content is 52.3 or 56 mL, the suffocation temperature is 150 to 153 degrees Celsius, and the amount of formaldehyde is optionally selected within the range of 29.1 to 38.9 mL.

For ease of reference, the results of the enthalpies and levels of the other naphthalene water-reducing agents, the index of the test, the order of the factors, and the optimum process parameters are summarized in Table I.1.3.

Table 1.1.3

Experimental number	NF				CU				FDN			
	A Condensation temperature (°C)	B Condensation Time (min)	C Formaldehyde Amount (g)	D Water (mL)	A Temperature (°C)	B Sulphuric acid Amount (g)	C Formaldehyde Amount (g)		A Temperature (°C)	B Acidity (%)	C Total acidity (%)	D Formaldehyde Amount (g)
1	95 – 100	60	0.6	30	140 – 145	1.4	0.4		120 – 125	24	31	0.7
2	105 – 110	150	0.8	50	150 – 155	1.6	0.6		140 – 150	26	29	0.9
3	115 – 120	300	1.2	80	160 – 165	1.8	0.8		155 – 160	28	33	1.2
Index	Fluent, cm				Fluent, cm				Fluent, cm			
Order	$D \to B \to C \to A$				$C \to B \to A$				$D \to A \to B \to C$			
Parameter	$A_1 B_3 C_3 D_1$				$A_3 BC$				$A_3 B_1 C_1 D_2$			
Company	Tsing University				Shanghai establish school				Zhanjiang medicine factory			

Appendix I Supplementary Examples

I.2 【Example I-2】Investigating the Compressive Strength of Low Temperature Curing on Wood-concrete Mixed Concrete Influences

Test purpose: A certain amount of wood calcium is inserted into the concrete. When the amount of cement is the same as the amount of water used, the slump is increased by about 10 cm; or the phase is maintaining the same slump can reduce the mixing water by 10% – 15% and increase the concrete compressive strength under standard curing conditions by about 15%. Experience shows that shuttle wood calcium concrete is not suitable for steam curing. Can it adapt to low temperature curing? It is a question that needs further experimental research. Wuxi Construction Engineering Company and other five units focused on the effects of cryogenic maintenance on the 28-day compressive strength of wood-cemented mixed button soil. Two kinds of cements were selected to maintain the same slump, and the "isolation variable method" was used to examine the strength variation of concrete under the three curing temperatures of 0 ℃, 10 ℃, and 20 ℃. According to the data obtained from the experiment, we arranged the experimental program according to the orthogonal design, examined the influence of curing temperature, wood calcium content and cement varieties on the compressive strength and their interactions.

I.2.1 Develop the Factor Level

Develop the factor level table I.2.1 and I.2.2 for both cases.

Table I.2.1 First condition

Level	Factor		
	A temperature(℃)	B Mu-Ca content(%)	C cement content(kg/m³)
1	0	0	PU-500,320
2	20	0.25	KUAZHA-500,350

Table I.2.2 Second condition

Level	Factor		
	A temperature(℃)	B Mu-Ca content(%)	C cement content(kg/m³)
1	10	0	PU-500,320
2	20	0.25	KUAZHA-500,350

I.2.2 Develop A Test Plan for Two Situations

Two $L_8(2^7)$ orthogonal tables were used to arrange the tests. The two tables consisted of 16 tests.

Orthogonal Design in Concrete Application

Since the standard cured concrete was commonly used for two different low temperature-curing conditions, 4 tests could be saved. Therefore, only It takes 12 trials to complete the two tables. The arrangement of the test plan and the method of calculation of range were the same as in Example 5.3.1. The results are shown in Tables I.2.3 and I.2.4. The last row in the two tables lists the approximate sum of the columns for analysis of variance.

Table I.2.3 First condition

Experimental number	A	B	$A \times B$	C	$A \times C$	$B \times C$		28-day Compressive strength(kg/cm^2)
	1	2	3	4	5	6	7	
1	1	1	1	1	1	1	1	304
2	1	1	1	2	2	2	2	193
3	1	2	2	1	1	2	1	345
4	1	2	2	2	2	1	2	197
5	2	1	2	1	2	1	2	458
6	2	1	2	2	1	2	1	359
7	2	2	1	1	2	2	2	556
8	2	2	1	2	1	1	1	407
K_1	1 040	1 314	1 460	1 664	1 416	1 366	1 416	
K_2	1 780	1 506	1 360	1 156	1 404	1 454	1 404	SUM:2 820
R	740	192	100	508	12	88	12	
S_i	68 450	4 608	1 250	32 258	18	968	18	

I.2.3 Analysis of Test Results
I.2.3.1 Intuitive Analysis

From the size of R, we can see the order of various factors and their interactions. For the first case:

Table I.2.4 Second condition

Experimental number	A	B	$A \times B$	C	$A \times C$	$B \times C$		28-day Compressive strength (kg/cm^2)
	1	2	3	4	5	6	7	
1	1	1	1	1	1	1	1	355
2	1	1	1	2	2	2	2	240
3	1	2	2	1	1	2	2	456
4	1	2	2	2	2	1	1	278
5	2	1	2	1	2	1	2	458
6	2	1	2	2	1	2	1	359

Continued Table I.2.4

Experimental number	A	B	A×B	C	A×C	B×C		28-day Compressive strength (kg/cm²)
	1	2	3	4	5	6	7	
7	2	2	1	1	2	2	1	556
8	2	2	1	2	1	1	1	407
K_1	1 320	1 412	1 558	1 825	1 577	1 498	1 548	
K_2	1 780	1 697	1 551	1 284	1 532	1 611	1 561	SUM:3 109
R	451	285	7	541	541	113	13	
S_t	25 425.125	10 153.125	6.125	36 585.125	36 585.125	1 596.125	21.125	

$A \to C \to B \to A \times B \to B \times C \to A \times C$, for the second case: $C \to A \to B \to B \times C \to A \times C \to A \times B$. It can be seen that the curing temperature is different and the primary and secondary orders are different. It is noteworthy that in the first case, wood-calcium-improved strengths are not as strong as in the second case; similarly, the difference between the strengths of the two different cements is smaller in the first case than in the second case. It is important to note that some interactions cannot be ignored.

I.2.3.2 Analysis of Variance

The results of the analysis of variance in both cases are listed in Table I.2.5 and Table I.2.6.

Table I.2.5 Variance analysis table of first condition

Variation source	Sum of squares	Freedom	Mean square	F value	Critical value
A	$S_A = 68\ 450$	1	68 450	3 802.8**	$F_{0.01}(1,2) = 98.5$
B	$S_B = 4\ 608$	1	4 608	256**	$F_{0.05}(1,2) = 18.5$
A×B	$S_{A \times B} = 1\ 250$	1	1 250	69.4*	
C	$S_C = 32\ 258$	1	32 258	1 792.1**	
A×C	$S_{A \times C} = 18$	1			
B×C	$S_{B \times C} = 968$	1	968	53.8*	
Error	$S_e = 18$	1			
e'	$S'_e = S_{A \times B} + S_e = 36$	2	18		
All	$S_T = 107\ 570$	7			

See Table I.2.5 and Table I.2.6 to see:

1. Common in both cases is that main effects A (curing temperature), B (calcium doping), and C (cement varieties) all have a particularly significant effect on strength; B × C has a significant effect on intensity or is particularly pronounced The effect of the combination of wood calcium content and cement variety has a significant or particularly significant effect on strength.

2. The main difference between the two cases is that when the curing temperature is close to 0 ℃, There was a significant interaction between A and B, ie the combination of curing temperature and wood calcium content had a significant effect on strength. At a curing temperature of approximately

10, there was no interaction between A and B, while A and C had significant interactions. That is, curing temperature and cement varieties together have a significant impact on strength.

Table I.2.6 Variance analysis table of second condition

Variation source	Sum of squares	Freedom	Mean square	F value	Critical value
A	$S_A = 25\,425.125$	1	25 425.125	1 866.1**	$F_{0.01}(1,2) = 98.5$
B	$S_B = 10\,153.125$	1	10 153.125	745.2**	$F_{0.05}(1,2) = 18.5$
$A \times B$	$S_{A \times B} = 6.125\,0$	1			
C	$S_C = 36\,585.125$	1	36 585.125	2 685.1**	
$A \times C$	$S_{A \times C} = 253.125$	1	253.125	18.6*	
$B \times C$	$S_{B \times C} = 1\,596.125$	1	1 596.125	117.1**	
Error	$S_e = 21.125$	1			
e'	$S'_e = S_{A \times B} + S_e = 27.25$		13.625		
All	$S_T = 74\,039.875$	7			

3. The test error in both cases is small, about 4 kg/cm²

I.3 【Example I-3】 Effect of the Dosage of the Molasses Plasticizer on Concrete and Easy-to-use

Test Purpose: To focus on the influence of the dosage of the molasses plasticizer on the workability of the concrete, the Second Engineering Bureau of the Guangdong Provincial Water Conservancy and Electric Power Bureau made a total of 64 test for slump with two kinds of cement, two kinds of water volumes, four kinds of water-cement ratios, and four Rinds of molasses plasticizers. Based on the data obtained from the test, 16 trials with ort hogonal design are used to analyze the test results.
Assessment indicators: slump.

I.3.1 Factors and Levels Are Shown in Table I.3.1

Table I.3.1 Factor level table

Factor	Level			
	1	2	3	4
A. Water cement ratio	0.50	0.58	0.66	0.74
B. Condensate	0	0.15	0.30	0.45
C. Type of cement	400#	500#		
D. Water volume	150	140		

I.3.2 Test Plan and Very Poor Needle Count

In this case, the water-cement ratio and the dosage are all four levels. Both the cement type and the water consumption are two levels. Select $L_{16}(4^2 \times 2^9)$ to arrange the test. The test plan and the range calculation results are shown in Table I.3.2.

Appendix I Supplementary Examples

Table I.3.2 $L_{16}(4^2 \times 2^9)$ Test results and calculation

Experimental number	A 1	B 2	C 3	D 4	5	6	7	8	9	10	11	Slump (cm)
1	1(0.50)	1(0)	1	1(140)	1	2	1	1	1	1	1	1.3
2	1	2(0.15)	1	1	1	1	2	2	2	2	2	3.3
3	1	3(0.30)	2	2(150)	2	2	1	1	2	2	2	15.4
4	1	4(0.45)	2	2	2	1	2	2	1	1	1	18.2
5	2(0.58)	1	1	2	2	1	2	2	2	2	2	7.9
6	2	2	1	2	2	2	1	1	2	1	1	11.9
7	2	3	2	1	2	1	2	2	1	1	1	11.4
8	2	4	2	1	2	2	1	1	1	2	2	12.2
9	3(0.66)	1	2	2	1	2	2	1	1	1	1	3.2
10	3	2	2	2	1	1	1	2	2	2	2	7.0
11	3	3	1	1	2	2	2	1	2	2	2	11.8
12	3	4	1	1	2	1	1	2	1	1	1	12.1
13	4(0.74)	1	2	2	1	1	2	1	2	2	1	5.4
14	4	2	2	2	1	2	1	2	1	1	2	6.3
15	4	3	1	1	2	1	2	1	1	1	1	3.4
16	4	4	1	1	2	2	1	2	2	2	2	5.0
$K_1(\bar{K}_1)$	38.2(9.6)	17.4(4.4)	56.3(7.0)	46.8(5.9)	63.8	66.0	67.1	68.7	67.7	67.8	72.0	SUM: 135.4
$K_2(\bar{K}_2)$	48.0(10.8)	28.5(7.1)	79.1(9.9)	88.6(11.1)	71.6	69.4	68.3	66.7	67.7	67.6	63.4	
$K_3(\bar{K}_3)$	34.1(8.5)	42.0(10.5)										
$K_4(\bar{K}_4)$	20.1(5.0)	47.5(11.9)										
R	22.9	30.1	22.8	41.8	7.8	3.4	1.2	2.0	0	0.2	8.6	

I.3.3 Analysis of Test Results

I.3.3.1 Intuitive Analysis

1) The order of the factors that influence the slump is: water consumption - molasses dosage - water-cement ratio, cement type.

2) Increase the water consumption, the slump increases significantly; in the case of increasing the molasses, the slump also increases significantly.

3) In the case of a certain water-cement ratio, water consumption, and dose, ordinary 500 cement has a higher slump than concrete 400.

4) In water ash, water consumption, dose, and cement varieties are certain
When the ratio is 0.58, the slump is the highest.

I.3.3.2 Analysis of Variance

1) Estimation of experimental error

The poor performance of the empty columns is the experimental error and they are averaged

$$\bar{R} = \frac{1}{7}(7.8 + \cdots + 8.6) = 3.3$$

here,

$$l = 7, n = 2, r = 8, d(n, l) = d(2, 7) = 1.18$$
$$\hat{\sigma} = \bar{R}/d(2,7)/\sqrt{r} = 3.3/1.18/\sqrt{8} = 0.989 = 1.0$$
$$f_e = \phi(2,7) = 0.90 l(n-1) = 0.9 \times 7 = 6.3 = 6$$

2) Test the significance of A, B, C, D

For A and B,

$$p = 4, r_A = r_B = 4, R_A = 22.9, R_B = 30.1$$
$$q_A = R_A/\hat{\sigma}/\sqrt{r_A} = 22.9/1/\sqrt{4} = 11.45^{**}$$
$$q_B = R_B/\hat{\sigma}/\sqrt{r_B} = 30.1/1/\sqrt{4} = 15.05^{**}$$

For C and D,

$$p = 4, r_C = r_D = 8, R_C = 22.8, R_D = 41.8$$
$$q_C = R_C/\hat{\sigma}/\sqrt{r_C} = 22.8/1/\sqrt{8} = 8.06^{**}$$
$$q_D = R_D/\hat{\sigma}/\sqrt{r_D} = 41.8/1/\sqrt{8} = 14.78^{**}$$
$$q_{0.01}(4,6) = 7.03, q_{0.01}(2,6) = 5.24$$

From the results of variance analysis, the amount of plasticizer, water consumption, water-cement ratio, and cement variety all had a significant effect on slump. The experimental error is about 1 cm and the accuracy is high.

I.3.3.3 Test the Significant Difference in Average Slump Between A and B Levels

For this example, $\bar{S}_e = \hat{\sigma}^2 = 1$, $f_e = 6$ for column 1 and 2, $m = 4$, $r = 4$; The average slumps of A and B are shown in Table I.3.2. From $q_{0.05}(m, f_e) = q_{0.05}(4,6) = 4.9$, calculate:

for A:

$$d_T = q_{0.05}(4,6)\sqrt{\frac{\bar{S}_e}{r}} = 4.9\sqrt{\frac{1}{4}} = 2.45$$

Appendix I Supplementary Examples

$d_{12} = |\bar{K}_1 - \bar{K}_2| = |9.6 - 10.8| = 1.2$

$d_{13} = |\bar{K}_1 - \bar{K}_3| = |9.6 - 8.5| = 1.1$

$d_{23} = |\bar{K}_2 - \bar{K}_3| = |10.8 - 8.5| = 2.3$

$d_{24} = |\bar{K}_2 - \bar{K}_4| = |10.8 - 5.0| = 5.8$

$d_{14} = |\bar{K}_1 - \bar{K}_4| = |9.6 - 5.0| = 4.6$

$d_{34} = |\bar{K}_3 - \bar{K}_4| = |8.5 - 5.0| = 3.5$

It can be seen that there is no significant difference between $\bar{K}_1$, $\bar{K}_2$, and $\bar{K}_3$, and there is a significant difference between $\bar{K}_4$ and other levels.

For B:

$d_{12} = |\bar{K}_1 - \bar{K}_2| = |4.4 - 7.1| = 2.7$

$d_{13} = |\bar{K}_1 - \bar{K}_3| = |4.4 - 10.5| = 6.1$

$d_{23} = |\bar{K}_2 - \bar{K}_3| = |7.1 - 10.5| = 3.4$

$d_{24} = |\bar{K}_2 - \bar{K}_4| = |7.1 - 11.9| = 4.5$

$d_{14} = |\bar{K}_1 - \bar{K}_4| = |4.4 - 11.9| = 7.5$

$d_{34} = |\bar{K}_3 - \bar{K}_4| = |10.5 - 11.9| = 1.4$

It can be seen that there is no significant difference between $\bar{K}_3$ and $\bar{K}_4$ and there are significant differences among the remaining levels.

From the results of multiple comparisons:

1. Under a certain cement type, water use, and certain agent settings, there is no significant difference in the average slump at any level between 0.5 and 0.66, and excessive water-cement ratio is significantly detrimental to the increase of slump.
2. Under certain cement varieties and water use and certain water-cement ratios, there is a significant difference in the average slump between undone, unbleached, 15%, and 0.30% molasses plasticizers; %, there is little effect on the slump height.

Discussion: The original author used the orthogonal table $L_{32}(4^5 \times 2^{16})$ on the basis of 64 times to do an analysis of variance (remove the cement variety factor.) Here, the author used the orthogonal table $L_{16}(4^2 \times 2^9)$ of the counterpart, and in the intuitive analysis Based on analysis of variance and multiple comparisons, not only the number of trials was reduced by 3/4, but also the results are more actual.

I.4 【Example I-4】 Examining the Effect of Mixing of Coal Ash, Fineness and Water-cement Ratio on Compressive Strength of Concrete

Test Objective 1 In the experimental study of fly ash quality, the different fineness of the fly ash, the effect of different dosing and different water-cement ratios on the concrete strength were investigated. A total of 288 groups of trials were arranged according to the "Orphan Transformation" method. After the test data were obtained, it was considered that the endogenous disorder might be due to poor test conditions, uneven materials, and incorrect conversions. Analyzing the test results and sawing only 48 test data, the influence rules of fly ash doping and fineness concrete strength were obtained.

Assessment indicators: 28-day and 90-day compressive strength.

I.4.1 Test Results and Test Results Table I.4.1 Shows the Turbidity Level

Table I.4.1 Factor level table

Factor	Level					
	1	2	3	4	5	6
A. Fly ash volume(%)	0	10	20	30	40	50
B. Fly ash size(%)	0-4	4-8	8-20			
C. Water cement ratio	0.5	0.7				

In this case, the admixture of fly ash is a six-level, the fineness is four levels, the water-cement ratio is two levels, and the orthogonally table of the counterpart is $L_{24}(6^1 \times 4^1 \times 2^3)$. The test results for the test and the calculation of the difference were listed in Table I.4.2 and I.4.3.

Table I.4.2 $L_{21}(6 \times 4^1 \times 2^3)$ Test results and calculation

Experimental number	A	B	C			Compressive strength(kg/cm^2)	
	1	2	3	4	5	28-day	90-day
1	1	1	3	2	2	323	416
2	1	2	1	1	1	358	375
3	1	3	2	3	3	206	268
4	2	1	2	1	2	204	285
5	2	2	3	3	1	314	422
6	2	3	1	2	3	371	404
7	3	1	1	3	1	192	357

Appendix I Supplementary Examples

Continued Table I.4.2

Experimental number	A	B	C			Compressive strength(kg/cm²)	
	1	2	3	4	5	28-day	90-day
8	3	2	2	2	3	181	239
9	3	3	3	1	2	299	404
10	4	1	1		3	350	435
11	4	2	2	1	2	170	219
12	4	3	3	3	1	148	212
13	5	1	3	2	3	125	170
14	5	2	1	3	2	144	201
15	5	3	2	2	1	327	368
16	5	1	2	1	1	225	304
17	5	2	3	2	3	116	143
18	5	3	1	1	2	118	174
19	5	3	2	1	2	215	306
20	5	4	2	2	1	189	262
21	6	1	1	2	2	95	136
22	6	2	1	1	1	94	138
23	6	3	2	2	1	153	221
24	6	4	2	1	2	164	227

Table I.4.3 Test result of calculate

		A Volume	B Size	C Water cement ratio			SUM
		1	2	3	4	5	
28-day Compressive strength	K_1	1 101	1 274	3 293	2 508	2 528	5 097
	K_2	1 058	1 435	1 804	2 589	2 569	
	K_3	966	1 268				
	K_4	822	1 120				
	K_5	633					
	K_6	512					
	R	589	315	1 489	81	41	

Continued Table I.4.3

		A Volume	B Size	C Water cement ratio			SUM
		1	2	3	4	5	
90-day Compressive strength	K_1	1 344	1 691	4 144	3 269	3 354	6 686
	K_2	1 422	1 727	2 542	3 417	3 332	
	K_3	1 270	1 739				
	K_4	1 043	1 529				
	K_5	885					
	K_6	722					
	R	700	210	1 602	148	22	

I.4.2 Analysis of Test Results

From Table I.4.3:

1) Regardless of the strength of 28-day or 90-day, the order of the primary and secondary impacts due to flooding is the same, namely, the ratio of water to cement→the amount of water→the fineness.
2) Regardless of 28-day or 90-day, milled fly ash has higher strength than the original fly ash. For the 28-day intensity > fineness with 4% - 8% more and for 90-day intensity, the fineness is less than 12%, and he does not make much difference.
3) The impact of fly ash content is calculated in Table 1.4.4. It can be seen that the 90-day intensity is lower than the 28-day intensity, indicating that fly ash can help it drop its activity at a later stage. When the amount is 20%, the strength does not decrease for 90-day.

Table I.4.4

Fly ash volume(%)		0	10	20	30	40	50
28-day Compressive strength	Average(kg/cm^2)	275	265	242	206	160	128
	Reduction rate(%)	0	-4	-12	-25	-42	-53
90-day Compressive strength	Average(kg/cm^2)	336	356	318	261	221	181
	Reduction rate(%)	0	6	-5	-22	-34	-46

I.5 【Example I-5】 To Investigate the Effect of Fly Ash Replacement Sand on the Strength and Impermeability of Concrete

Test purposes: Guangxi Dahua Hydropower Station Project Command In order to make full use of artificial rolling sand and improve its configuration, in addition to the use of certain fine sand in the grit after sieving, some fine sand is used instead of fine ash. Sand to further improves the ease of use and impermeability of concrete.

Evaluation index: 28-day compressive strength and impermeability labels.

I.5.1 Test Plan and Test Results

I.5.1.1 Factors and Levels

Factors and levels are listed in Table I.5.1

Table I.5.1 Factor level table

Level	Factor			
	A. Water cement ratio	B. Fine sand content(%)	C. Fly ash sand content(%)	D. Fly ash cement content(%)
1	0.5	25	0	60
2	0.6	30	9	0
3	0.7	35	6	20
4	0.8	40	3	40

I.5.1.2 Test Plan and Range Calculations

Test plan and range calculations are listed in Table I.5.2.

Table I.5.2 $L_{16}(4^5)$ Test scheme and calculation results

Experimental number	A	B	C	D		28-day Compressive strength	Impermeability labeling
	1	2	3	4	5		
1	1(0.5)	1(25)	1(0)	1(60)	1	219	8
2	1	2(30)	2(9)	2(0)	2	434	22①
3	1	3(35)	3(6)	3(20)	3	402	12
4	1	4(40)	4(3)	4(40)	4	320	12
5	2(0.6)	1	2	3	4	351	36①
6	2	2	1	4	3	237	5
7	2	3	4	1	2	189	6
8	2	4	3	2	1	354	11

Continued Table I.5.2

Experimental number	A 1	B 2	C 3	D 4	5	28-day Compressive strength	Impermeability labeling
9	3(0.70)	1	3	4	2	218	5
10	3	2	4	3	1	243	11
11	3	3	1	2	4	276	10
12	3	4	2	1	3	172	11
13	4(0.80)	1	4	2	3	232	3
14	4	2	3	1	4	136	3
15	4	3	2	4	1	187	11
16	4	41	1	3	2	168	4
$K_1(\bar{K}_1)$	1 375 (343.8)	1 020 (255)	900 (225)	716 (179)	1 003	SUM:4138	SUM:170
$K_2(\bar{K}_2)$	1 131 (282.8)	1 050 (262.5)	1 144 (286)	1 296 (324)	1 009		
$K_3(\bar{K}_3)$	909 (227.3)	1 054 (263.5)	1 110 (277.5)	1 164 (240.5)	1 043		
$K_4(\bar{K}_4)$	723 (180.8)	1 014 (253.5)	934 (246)	926	1 083		
R	625	40	244	580	80		
$K_1(\bar{K}_1)$	54 (13.5)	52 (13)	28 (7)	27 (6.8)	41		
$K_2(\bar{K}_2)$	58 (14.5)	42 (10.5)	80 (20)	46 (11.5)	36		
$K_3(\bar{K}_3)$	37 (9.3)	38 (9.5)	31 (7.8)	63 (15.8)	32		
$K_4(\bar{K}_4)$	21 (5.3)	38 (9.5)	31 (7.8)	34 (8.5)	61		
R	37	14	52	36	29		

I.5.2 Analysis of Test Results

I.5.2.1 Direct Analysis

1. The main cause of the impact strength is water-cement ratio and fly ash cement ratio. The secondary factor is the percent string of fly ash replacement sand. The main factor influencing the impervious marking is the percentage of fly ash replacement sand, and the secondary factor is the ratio of water to cement and the cement ratio of fly ash. Fine sand content has little effect on strength and impermeability labels and is within test error. It can be seen that the assessment

Appendix I Supplementary Examples

indicators are different, and the influences on the order of the first and second precepts of Chong are not the same.

2. The compressive strength decreases with the increase of the fly ash cement ratio. When the cement ratio exceeds 20%, the strength decreases significantly.
3. Compressive strength increases with the percentage of fly ash replacement sand, and as the percentage of contemporary sand increases from 6% to 9%, there is little increase in strength.
4. Impervious labels increase with the percentage of fly ash replacement sand. When the percentage of sand replacement is below 6%, the impermeability mark is not much improved; when the percentage of contemporary sand exceeds 6%, the impermeability mark is significantly increased.

I.5.2.2 Analysis of Variance

For 28-day intensity:

1) Estimated test error

Columns 2 and 5 are combined to jointly estimate the test error. Average difference

$$\overline{R} = 0.5 \times (40 + 80) = 60$$

in this example $l = 2$, $n = 4$, $r = 4$ find $d(n, l)$ table $d(4,2) = 2.15$, then

$$\hat{\sigma} = \overline{R}/d(4,2)/\sqrt{r} = 60/2.15/2 = 14$$
$$f_e = \phi(4,2) = 6$$

2) Examination of the significance of A, C, D

$R_A = 625$, $R_C = 244$, $R_D = 580$, $m = 4$, $r_A = r_C = r_D = 4$, then

$$q_A = 624/14/2 = 23.32^{**}$$
$$q_C = 224/14/2 = 8.0^{**}$$
$$q_D = 580/14/2 = 20.71^{**}$$
$$q_{0.05}(m, f_e) = q_{0.05}(4,6) = 4.9$$
$$q_{0.01}(4,6) = 7.03$$

3) Multiple comparison of factor C

$m = 4$, $r = 4$, $\overline{S}_e = \hat{\sigma}^2 = 14^2 = 196$, $f_e = 6$

$\overline{K}_1 = 225$, $\overline{K}_2 = 286$, $\overline{K}_3 = 277.5$, $\overline{K}_4 = 246$,

$q_{0.05}(4,6) = 4.9$

$$d_T = q_{0.05}(4,6)\sqrt{\frac{S_e}{r}} = 4.9\sqrt{\frac{196}{4}} = 34.3$$

Calculate

$$d_{12} = |\overline{K}_1 - \overline{K}_2| = |225 - 286| = 61$$
$$d_{13} = |\overline{K}_1 - \overline{K}_3| = |225 - 277.5| = 52.5$$
$$d_{14} = |\overline{K}_1 - \overline{K}_4| = |225 - 246| = 21$$
$$d_{23} = |\overline{K}_2 - \overline{K}_3| = |286 - 277.5| = 8.5$$
$$d_{24} = |\overline{K}_2 - \overline{K}_4| = |286 - 246| = 40$$
$$d_{34} = |\overline{K}_3 - \overline{K}_4| = |277.5 - 246| = 31.5$$

Analysis of the variance of the strength index shows that the water pressure ratio, fly ash cement ratio, and sand replacement percentage have a particularly significant effect on the 28-day compressive strength. The percentage of substitution sand percentage of 6% is significantly different from the intensity of non-substitution, but it is not significantly different from the percentage of substitution sand of 9%. The compressive strength test error of $\hat{\sigma} = 14$ kg/cm^2 is less accurate.

Anti-seepage marking:

1) Estimated test error

$$\bar{R} = \frac{1}{2} \times (R_2 + R_5) = \frac{1}{2}(14 + 29) = 21.5$$

$l = 2, n = 4, r = 4$

$d(n,l) = d(4,2) = 2.15$

then $\hat{\sigma} = 21.5/2.15/\sqrt{4} = 5$

$f_e = \phi(4,2) = 6$

2) Test the significance of $A, C,$ and D

$R_A = 37, R_C = 52, R_D = 36, m = 4$

$r_A = r_C = r_D = 4$

$q_A = 37/5/2 = 3.7$

$q_C = 52/5/2 = 5.2^*$

$q_D = 36/5/2 = 3.7$

$q_{0.05}(4,6) = 4.9, q_{0.1}(4,6) = 4.07$

3) Multiple comparison of factor C

$m = 4, r = 4, S_e = \hat{\sigma}^2 = 5^2 = 25, f_e = 6$

$\bar{K}_1 = 7, \bar{K}_2 = 20, \bar{K}_3 = 7.75, \bar{K}_4 = 7.75$

$d_T = q_{0.05}(4,6)\sqrt{\frac{S_e}{r}} = 4.9\sqrt{\frac{25}{4}} = 12.25$

Calculate

$d_{12} = |\bar{K}_1 - \bar{K}_2| = |7 - 20| = 13$

$d_{13} = |\bar{K}_1 - \bar{K}_3| = |7 - 7.75| = 0.75$

$d_{14} = |\bar{K}_1 - \bar{K}_4| = |7 - 7.75| = 0.75$

$d_{23} = |\bar{K}_2 - \bar{K}_3| = |20 - 7.75| = 12.25$

$d_{24} = |\bar{K}_2 - \bar{K}_4| = |20 - 7.75| = 12.25$

$d_{34} = |\bar{K}_3 - \bar{K}_4| = |7.75 - 7.75| = 0$

The variance analysis of the impermeability label shows that the percent of fly ash replacement sand has a significant effect on the anti-seepage mark, and the water-cement ratio and the fly ash cement ratio have no effect. When fly ash is used for 6% of sand, its impervious index is not significantly different from that of non-substitution. The anti-dosage label of 9% of the replacement sand is only slightly different from the anti-seepage label of 6% of replacement sand. Test error of smpermeability label $\sigma = 5$ kg/cm^2, poor precision.

To sum up, in order to prepare a concrete with a compressive strength of 250 – 300 kg/cm^2 and an impermeability rating greater than 8 kg/cm^2, the ratio of water to cement is not greater than 0.60, the percentage of fly ash substitute sand is not less than 6%, and the cement substitution rate is 20% is appropriate.

I.6 【Example I-6】 Ring Crack Test of Cement Type and Expansion Agent

Objective of the experiment: The Yangtze River Institute of Water Resources and Hydropower Research uses three different types of cement and incorporates two different types of expansive agents to perform the cement paste ring crack resistance test. The purpose of the test was to compare the crack resistance of the cement and the expansion agent. The test evaluation index is the time (in days) when the ring crack occurred, and the later the crack appeared, the better the crack resistance of the material.

I.6.1 Test Plan and Calculation of Results

According to the test data, develop the factor level table I.6.1.

Table I.6.1 Factor level table

Level	Factor		
	1	2	3
A. Cement	Huaxin-600$^{\#}$	Huaxin-500$^{\#}$	Huaxin-500$^{\#}$ low swell
B. Swell	0	15% * E-12	10% * E-40

Arrange the test with the table $L_9(3^4)$ and leave two columns free for estimating the test error. The test protocol and test results are listed in Table I.6.2.

Table I.6.2 $L_1(3^4)$ Test scheme and calculation results

Experimental number	A	B			Fractured days
	1	2	3	4	
1	1	1(0)	1	1	3
2	1	2(15% E-12)	2	2	12
3	1	3(10% E-40)	3	3	6
4	2	1(0)	2	1	3
5	2	2(15% E-12)	3	2	14
6	2	3(10% E-40)	1	3	9
7	3	1(0)	3	2	19

Continued Table I.6.2

Experimental number	A	B			Fractured days
	1	2	3	4	
8	3	2(15% E-12)	1	3	28
9	3	3(10% E-40)	2	1	24
K_1	21	25	40	41	SUM:118
K_2	26	54	36	40	
K_3	71	39	36	37	
R	50	29	1	4	

I.6.2 Analysis of Test Results

The group with the longest crack appeared directly was No. 8 test. The period of cracks was 28-day.

Calculating the calculation result of the range difference is shown below in Table 1.6.2. From the magnitude of the range R, it can be seen that column A is the largest and B is the second, that is, the effect of the cement variety on the cracking time of the ring is more important than the expansion agent. In the A column, the A_3 is the best, that is, the No. 500 low-thermal micro-expansion cement has better crack resistance than the other two cements; the B column is best listed in B_2, that is, the 15% E-12 type expansion agent has good crack resistance. The preferred combination is A_3B_2, which is the 8th trial number.

The sum of the extremes of the two empty columns is $(1+4) = 5$, the average is $5/2 = 2.5$, and the number is $2.5/3 = 0.83$, which means that the estimated value of the experimental error in this example is about 1 day.

This example is a full-scale trial of two-factor cross grouping. We use orthogonal designs for the original data to make the conclusions clearer and clearer; at the same time, we give an estimate of the experimental error.

I.7 【Example I-7】 Examining the Effect of Fluidizer Dosage and Method of Addition (Pre-additive and Post-addition) on the Slump of Flowing Concrete

The so-called first addition is to add the fluidizing agent together with the mixing water to the cement, sand, and stone while stirring; after the addition of the reference concrete, after a certain time interval, the fluidizing agent is added for secondary stirring. The concrete slump when it is poured is more than 20 cm. This is a new type of concrete that prevents loss of twist and is suitable for pumping.

The purpose of the experiment, the doping method of fluidizing agent and the mixing of s are two important factors for the preparation of flow concrete. In order to clear the law, the Women's Medical Institute Highway Institute and the Chinese People's Liberation Army's 00069 Troops were set at 0.1% - 0.6% Jiangdu MF water reducer was subjected to a first and a subsequent addition test to find out the difference between the first addition and the latter addition to determine the optimum process conditions.

Assessment Index: Slump during casting (in this test, it refers to slump at 10 minutes or 60 minutes after mixing).

I.7.1 Test Plan and Test Results

The factors and levels are listed in Table I.7.1.

Table I.7.1 Factor level table

Level	Factor					
	1	2	3	4	5	6
A. content(%)	0.2	0.3	0.4	0.5	0.6	0.1
B. Type	First add	Post add				
C. Time	10	60				

The dosage of this example is six levels. The mixing method and casting time are two levels. The orthogonal table of the counterpart is $L_{12}(6^1 \times 2^2)$, and the experimental scheme and range calculation results are listed in Table 1.7.2. For ease of comparison, plots of changes in measured slump during the first addition and the next addition casting under different doping are shown in Figure I.7.1.

Table I.7.2 $L_{12}(6^1 \times 2^3)$ Test scheme and calculation results

Experimental number	A	B	C	Slump	
	1	2	3	Mixing	Pouring
1	1(0.2)	1(First add)	1(10)	9.8	3.1
2	2(0.3)	1	2(60)	15.1	4.1
3	1	2(Post add)	2	7.0	10.5
4	2	2	1	4.6	20.2
5	3(0.4)	1	2	20.5	4.0
6	4(0.5)	1	1	20.2	11.0
7	3	2	1	5.4	20.2
8	4	2	2	4.5	21.6
9	5(0.6)	1	1	21.1	8.2
10	6(0.1)	1	2	13.3	2.4
11	5	2	2	4.6	21.3

Continued Table I.7.2

Experimental number	A	B	C	Slump	
	1	2	3	Mixing	Pouring
12	6	2	1	8.4	6.0
K_1	13.6	32.8	68.7	SUM:132.6	
K_2	24.3	99.8	63.9		
K_3	24.2				
K_4	32.6				
K_5	29.5				
K_6	8.4				
R	24.2	67.0			
$\overline{K_1}$	7.8	5.5	11.5		
$\overline{K_2}$	12.2	16.6	10.7		
$\overline{K_3}$	12.1				
$\overline{K_4}$	16.8				
$\overline{K_5}$	14.8				
$\overline{K_6}$	4.2				

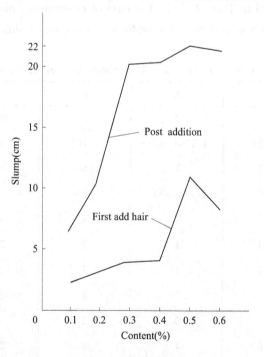

Figure I.7.1 *Relationship between dosage and slump*

I.7.2 Analysis of Test Results

See Table I.7.2 and Figure I.7.1:

1. The doping method is the main factor affecting the slump of the casting mold, the dose is a secondary factor, and the casting mold time has little effect.
2. Compared with the previous addition method, the former slump is 11.0 cm larger than the latter average when cast, which indicates that the post-adding method can fully exert the effect of the fluidizing agent.
3. After the addition, if the slump is over 20 cm when pouring the mold, the MF content of Jiangdu is between 0.4 and 0.6.
4. The best process condition is $A_4B_2C_2$ (or C_1), ie 0.5% dosing, post-addition, and die time 10 or 60 minutes. This is the test condition of Test No. 8.

I.8 【Example I-8】Early Strength and High Strength Concrete Test

Purpose of the test: The concrete with a certain amount of Mt-150 admixture can be used to prepare tough concrete. Through the test, it is ascertained that the Mt-150 dosage, the type of cement, and the effect of the application on the strength of the early strength reinforced concrete are selected and the best technology is selected condition.

Assessment indicators: 7-day and 28-day compressive strength.

I.8.1 Factors and Levels

Factors and levels see Table I.8.1 for factors and levels.

Table I.8.1 Factor level table

Level	Factor		
	A. cement volume (kg/m^3)	B. MT-150 volume (%)	C. Cement type
1	400	0	Putong
2	500	1	Kuaiying
3	600		

I.8.2 Test Plan and Test Results

This example A is three levels, B and C are two levels, and the orthogonal table of the counterpart is $L_{12}(3^1 \times 2^4)$. The test plan and test results are shown in Table I.8.2.

Table I.8.2 $L_{12}(3^1 \times 2^4)$ Test scheme and calculation results

Experimental number	A	B	C			7-day Compressive strength (kg/cm²)	28-day Compressive strength (kg/cm²)	Attach	
	1	2	3	4	5			Water cement ratio(%)	slump (cm)
1	2	1	1	1	1	593	802	37.0	8.8
2	2	2	2	2	2	915	1 009	27.4	9.5
3	2	3	3	3	3	662	752	38.0	6.0
4	2	4	4	4	4	923	1 078	29.6	7.8
5	1	1	2	3	4	404	635	42.5	9.5
6	1	2	1	4	3	824	976	31.5	9.8
7	1	3	4	1	2	581	685	43.8	6.0
8	1	4	3	2	1	845	968	34.3	7.8
9	3	1	3	4	2	752	850	33.3	7.8
10	3	2	4	3	1	983	1 150	25.3	7.1
11	3	3	1	2	4	690	805	34.2	8.1
12	3	41	1	3	2	1 072	1 210	26.7	8.2
K_1	3 093	3 718	4 507	4 677	4 685				
K_2	2 690	5 562	4 773	4 603	1 595	SUM:9 280 SUM:10 920			
K_3	3 497								
R	807	1 844	266	74	90				
K_1	3 641	4 529	5 422	5 533	5 403				
K_2	3 264	6 391	5 498	5 387	5 517				
K_3	4 015								
R	751	1 862	76	146	114				

Appendix I Supplementary Examples

I.8.3 Analyze the Test Results

1. Directly see the 12th and 28th days of the 12th test. The strongest squats are the most severe, and the combination condition is $A_3B_2C_2$.
2. Calculated Range The calculated results are listed below in Table I.7.2. From the size of R, it can be seen that Mt-150 dose is the main factor regardless of 7-day intensity or 28-day intensity, cement content is the second factor, and cement strength has a certain influence on 7-day strength; and for 28-day strength The impact is small and within the test error. From the size of K_i, the better combination condition is $A_3B_2C_2$, which is exactly the same as the direct observation condition.

In order to obtain early-strength high-strength concrete with strength greater than 1 000 kg/cm², 600 g/m³ of fast-hardening cement was used, and 1% Mt-150 admixture was added.

I.9 【Example I-9】Lightweight Aggregate Concrete Mix Ratio Test

Purpose of the test: The bridge science research institute of the Ministry of Railways Bridge Engineering Bureau passed the test. The proportion of lightweight aggregate concrete with a designation of 600 kg/cm² was selected. Evaluation indicators: 28-day compressive strength.

I.9.1 Test Plan and Results the Formulation

Factors and levels are shown in Table I.9.1.

Table I.9.1 Factor level table

Level	Factor			
	A. Water cement ratio	B. Fine sand content	C. Bundle	D. Ceramic species
1	0.28	0.55	1.25	Red 1#
2	0.32	0.75	1.20	Red 2#
3	0.3	0.95	1.15	
4	0.40	1.15	1.10	

The $L_{16}(4^3 \times 2^6)$ orthogonal table has nine columns. It can arrange three four-level factors and up to six two-level factors. This case has three four-level factors and one two-level factor. The test plan and results are shown in Table I.9.2.

Orthogonal Design in Concrete Application

Table I.9.2 $L_{16}(4^3 \times 2^6)$ Test results and calculation

Experimental number	A. Water cement ratio (1)	B. Fine sand content (2)	C. Bundle (3)	D. Ceramic species (4)	5	6	7	8	9	28-day Compressive strength (kg/cm^2)
1	1(0.28)	1(0.55)	1(1.25)	1(Red 1)	1	1	1	1	1	587
2	1	2(0.75)	2(1.20)	1	1	2	2	2	2	592
3	1	3(0.95)	3(1.15)	2	2	1	1	2	2	619
4	1	4(1.15)	4(1.10)	2	2	2	2	1	1	560
5	2(0.32)	1	2	2	2	1	2	1	2	584
6	2	2	1	2	2	2	1	2	1	611
7	2	3	4	1	1	1	2	2	1	594
8	2	4	3	1	1	2	1	1	2	544
9	3(0.36)	1	3	2	2	2	2	2	1	496
10	3	2	4	2	2	1	1	1	2	508
11	3	3	1	1	1	2	2	1	2	540
12	3	4	2	1	1	1	1	2	1	558
13	4(0.40)	1	4	1	1	2	1	2	2	528
14	4	2	3	1	1	1	2	1	1	512
15	4	3	2	2	2	2	1	1	1	544
16	4	4	1	2	2	1	2	2	2	480
K_1	2 358	2 195	2 218	4 345	4 455	4 442	4 499	4 379	4 462	SUM:8 857
K_2	2 333	2 223	2 278	4 512	4 402	4 415	4 358	4 478	4 395	
K_3	2 102	2 297	2 171							
K_4	2 064	2 142	2 190							
R	294	155	107	167	53	27	141	99	67	

Note: This example $L_{16}(4^3 \times 2^6)$ is equivalent to Table 12 $L_{16}(4^3 \times 2^6)$ in appendix Ⅲ.

Appendix I Supplementary Examples

I.9.2 Analysis of Test Results

The highest strength of the No. 3 test is 619 kg/cm² directly, and the strength of the No. 6 test number is the next highest, and its value is 611 kg/cm². Their combination conditions are $A_1B_3C_3D_2$ and $A_2B_2C_1D_2$.

Calculating the calculation result of the range difference is shown below in Table 1.9.2. The order of intensity of the factors is $A \to D$, $B \to C$, which means that the ratio of water to cement is the main factor affecting the strength. The type of ceramist and sand-cement ratio are secondary factors, and the influence of the package factor is smaller. From the size of K_i, better combination conditions are $A_1B_3C_2D_2$. However, it is not included in 16 trials. To determine the mix ratio, further analysis of variance is required.

I.9.2.1 Estimated Test Error

$L_{16}(4^3 \times 2^6)$ has no arrangement of empty columns and can be used to break experimental errors and average them

$$R = 0.2 \times (53 + 27 + 141 + 99 + 67) = 77.4$$

here $l = 5$, $n = 2$, $r = 8$, check $d(n, l)$ table we can get

$$d(n, l) = d(2, 5) = 1.19$$

then

$$\hat{\sigma} = \overline{R}/d(2,5)/\sqrt{r} = 77.4/1.19/\sqrt{8} = 23.0$$
$$f_e = \phi(2,5) = 4.6 \approx 5$$

I.9.2.2 Examining the Significance of A, B, C

$R_A = 294$, $r_A = m = 4$

$q_A = 294/23/\sqrt{4} = 6.39^*$

$R_B = 155$, $r_B = m = 4$

$q_B = 155/23/\sqrt{4} = 3.37$

$R_C = 107$, $r_C = m = 4$

$q_C = 107/23/\sqrt{4} = 2.33$

$R_D = 167$, $r_D = m = 4$

$q_D = 167/23/\sqrt{4} = 2.57$

$q_{0.05}(4,5) = 5.22$, $q_{0.01}(4,5) = 7.80$

$q_{0.05}(2,5) = 3.64$, $q_{0.1}(2,5) = 2.85$

From the analysis of variance, it can be seen that except for the water-cement ratio (A), which has a significant effect on the strength, other factors do not have any effect on the strength.

I.9.2.3 Comparison of Level 1 and Level 2, 3, and 4 of Factor A

$m = 4$, $r = 4$, $\overline{S}_e = \hat{\sigma}^2 = 23^2 = 529$, $f_e = 5$

$\overline{K}_1 = 589.5$, $\overline{K}_2 = 583.3$, $\overline{K}_3 = 525.5$, $\overline{K}_4 = 516$

then

$$d_T = q_{0.05}(4,5)\sqrt{\frac{S_e}{r_A}} = 5.22\sqrt{\frac{529}{4}} = 60$$

Calculate

$$d_{12} = |\overline{K}_1 - \overline{K}_2| = |589.5 - 583.3| = 6.25$$
$$d_{13} = |\overline{K}_1 - \overline{K}_3| = |589.5 - 525.55| = 64$$
$$d_{14} = |\overline{K}_1 - \overline{K}_4| = |589.5 - 516| = 73.5$$

It can be seen that there is no significant difference between level 1 and level 2, and there is a significant difference between level 1 and levels 3 and 4.

Therefore, in order to obtain the No. 600 lightweight aggregate concrete, the water-cement ratio should be less than 0.32, and the other factors should be better, which can be selected in the test range.

I.10 〖Example I-10〗 Design of Steel Fiber Concrete Mixture Ratio

Purpose of the test: In the test study of the anti-abrasion material at the bottom of the Sanmenxia project of the Ministry of Water Resources, Tianjin Survey and Design Institute and the First Engineering Bureau, steel fiber reinforced concrete is proposed as a wear-resistant and impact-resistant material. In the preliminary selection of anti-wear materials, Li seeks steel fiber reinforced concrete for 28-day compressive strength of 800 kg/cm^2 (10 cm × 10 cm × 10 cm cubic specimen), bending strength of 100 kg/cm^2, slump of 10 to 12 cm, and Good workability. Try to determine the main parameters of its mix ratio.

The raw materials in the test are as follows:

Cement: Jingmen 525 cinematic cement.

Coarse fine aggregate: gravel with a maximum particle size of 20 mm, compacted bulk density of 1 750 kg/m^3; river sand, fineness modulus of 2.84.

Water reducing agent FDN dosage 0.5%.

Steel fiber: diameter 0.5 mm, length 40 mm (with hook).

Assessment indicators: 28-day compressive strength, bending strength and slump.

Insole and level are listed in Table I.10.1.

Table I.10.1 Factor level table

Level	Factor		
	A. Steel fiber	B. Pebble content(kg/m^3)	C. Water cement ratio
1	0	850	0.28
2	0.8	750	0.32
3	1.3		
4	1.8		

The experimental calculations and the extreme calculations of the test results are listed in Table I.10.2.

The relationship between average slump, 28-day compressive strength, flexural strength, and steel fiber content (percentage of concrete volume) are shown in Figure I.10.1 (a), (b), respectively. (c).

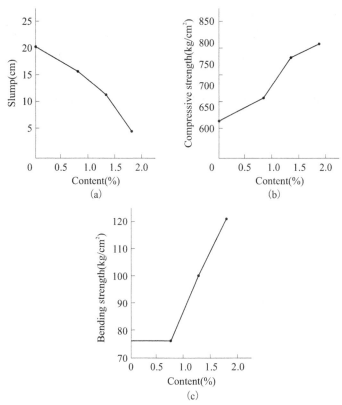

Figure I.10.1 *The relationship between the Slump(a), Compressive strength(b), Bending strength(c) and Steel fiber content*

Concrete specimens are machined by mixing, shake table forming, and water conservation. The compressive strength is 10 cm × 10 cm × 10 cm cube specimen strength results.

Visual analysis results show:

1. Steel fiber blending is the main factor affecting the slump and flexural strength, and the effect of stone cargo and water-cement ratio is within the margin of error.
2. Under certain conditions, slump decreases with the increase of steel fiber admixture. For every 0.5% (volume percentage) of steel fiber, the average slump decreases by about 5 cm.
3. When the content of steel fiber is less than 0.8% (volume percentage), the bending strength is basically not improved; only when the content of the steel fiber is increased to more than 1%, the bending strength is obviously lifted, and the maximum amount is added. At 1.8%, the average flexural strength was about 1.6 times higher than when it was not blended.
4. The main factor affecting the compressive strength is the content of steel fiber, the second factor is the ratio of water to cement, and the amount of stone has no effect. The compressive strength increases with the increase of steel fiber content. When the maximum content is 1.8% (volume

percentage), the strength is about 20%.
5. The combined condition for meeting the slump and bending strength requirements for $A_3 B_0 C_3$ to reach the syncope compressive strength is the $A_4 B_0 C_1$; Conbined slump, compressive strength, and bending strength indexes, taking into account that when the steel fiber is doped with 1.8%, the fibers tend to form clusters during the mixing process, when the dosage of stone is B_2, the workability is good. Therefore, the $A_3 B_2 C_1$ parameters for the preliminary determination of the mix ratio are: This is the combined condition of the No. 6 test, that is, the steel fiber. The amount is 1.3%, the amount of stone is 750 kg/m^3, and the water-cement ratio is 0.28. The measured slump was 11.6 cm, the compressive strength was 804 kg/cm^2 for 28-day, and the flexural strength was 107.7 kg/cm^2.

Appendix I Supplementary Examples

Table I.10.2 $L_{16}(4^3 \times 2^6)$ Test results and calculation

Experimental number	A	B	C			28-day Compressive strength (kg/cm^2)	28-day Flexural rigidity (kg/cm^2)	Calculation of bulk density γ (kg/m^3)
	1	2	3	4	5	Slump (cm)		
1	1(0)	1(850)	1(0.28)	1	1	21.4	690	74.8	2 400
2	1	2(750)	2(0.32)	2	2	19.8	662	77.5	2 400
3	2(0.8)	1	1	2	2	16.0	730	77.1	2 410
4	2	2	2	1	1	16.4	707	76.7	2 410
5	3(1.3)	1	2	1	2	16.4	707	76.7	2 410
6	3	2	1	2	1	11.6	804	107.7	2 435
7	4(1.8)	1	2	2	1	5.0	807	124.0	2 470
8	4	2	1	1	2	4.9	836	120.6	2 470
						SUM:8 857	SUM:6 018	SUM:754.4

		A	B	C	4	5
Slump (cm)	$K_1(\overline{K_1})$	41.2	54(13.5)	53.9(13.5)	54.3(13.6)	54.4(13.6)
	$K_2(\overline{K_2})$	32.4	52.7(13.2)	52.8(13.2)	52.4(13.1)	52.3(13.1)
	$K_3(\overline{K_3})$	23.1				
	$K_4(\overline{K_4})$	9.9				
	R		1.3	1.1	1.9	2.1
Compressive strength	$K_1(\overline{K_1})$	2 352	3 009(752)	3 060(765)	3 015(754)	3 008(752)
	$K_2(\overline{K_2})$	1 437	3 009(752)	2 958(740)	3 003(751)	3 010(753)
	$K_3(\overline{K_3})$	1 586				
	$K_4(\overline{K_4})$	291				
	R		0	102	12	2
Flexural rigidity	$K_1(\overline{K_1})$	152.3(76.2)	371.9(93)	380.2(95)	368.1(92)	383.2(95.8)
	$K_2(\overline{K_2})$	153.8(76.9)	382.5(95.6)	374.2(93.6)	386(96.6)	371.2(92.8)
	$K_3(\overline{K_3})$	203.7(101.8)				
	$K_4(\overline{K_4})$	244.6(122.3)				
	R	92.3	10.6	6.0	18.2	12.0

Appendix II
Whole Book Case Index

Category	No.	Project	Characteristic	Orthogonal table	Case number	Page number
Cement	1	High strength Portland cement	① Optimization of mineral composition of high strength cement clinker ② Orthogonalddesign of compound factars	$L_{24}(3^1 \times 4^1 \times 2^4)$	2-4-1	
	2	Sulfate cement	① Optimization of gypsum composition of high strength cement clinker ② t-stablishment of two variable regression equation	$L_9(3^4)$	6.2.1	
Water reducer	3	High efficiency water reducing agent	①Optimizing the technological parameters of DH-3 water reducing agent and summing up the production technology of domestic water reducer. ② Sinplified ANOVA for repeated trials	$L_9(3^4)$	I-1	
	4	Three kinds of DH products	The orthogonal test is applied to the cross group test. Empty colums without snteraotion are shown to be experimented errors	$L_9(3^4)$	3-2-2	
					4-2-2	
			The simplified variance ana lysis is consistent with the commonly used variance analysis		4-3-1	
			Multiple comparison			
	5	Molasses plasticizer	①Optimal dosage ② Choose an appropriate orthogonal table ③Multiple comparison	$L_{16}(4^2 \times 2^9)$	1-3	

Appendix II Whole Book Case Index

Continued

Category	No.	Project	Characteristic	Orthogonal table	Case number	Page number
Water reducer	6	Low temperature curing of calcium concrete mixed with wood	Orthogonal design can be used to analyze the interaztion in stead of the isolated varable method	$L_8(2^7)$	I-2	
	7	Low temperature curing of DH and calcium concrete	The concept of interaction and intuitionistic judgment	$L_4(2^3)$	5-1-1	
	8	Effect of MF and NNO on vibration technology in concrete	①Investigation of enhancement effect, ②distinction between interaction and test error.	$L_4(2^3)$	5-2-1	
	9	The optimal dosage of each component of water reducing retarder	① Selection of the optimal dosage of 3FG components ② Multindex analysis method-Efficacy coefficient methed	$L_9(3^4)$	2-5-1	
Fly ash	10	Study on the quality standard of fly ash	① Nearly 300 experiments have been made with the method of "isolated variable" the reasults are irregular. Orthaganal design uses 24 of them to find out the rules.	$L_{24}(6^1 \times 4^1 \times 2^3)$	I-4	
	11	Anti freezing of fly ash aerated concrete	① Optimization of fineness and content of fly ash ②Activity level and its arrangement method	$L_{12}(3^1 \times 2^4)$	2-3-1	
	12	The former shareholders and impermeability of fly ash	①Multiple comparison of the percentage of preferred sand generation ②Multiple comparison	$L_{16}(4^5)$	I-5	
Expansive agent	13	Round ring crack resistance test of expansive agent	①Preferred varieties and expansive agents ②Multiple comparison	$L_9(3^4)$	I-6	

· 149 ·

Continued

Category	No.	Project	Characteristic	Orthogonal table	Case number	Page number
Selection of match ratio Expansive agent	14	Optimization ratio of large Luo Du concrete with strength not lower than cement hardening index	①If a comprehensive test is required 125 times, the orthogoned design only needs 17 times. ②The multi index analysis method — amprehensine balance	$L_9(3^4)$ & $L_8(2^7)$	2-5-1	
	15	Multiple regression equation of concrete collapse	① Multiple regression analysis method was used to ② establish the three element regression equation of slump to ash water ratio and FDN content.	$L_9(3^4)$	6-3-1	
	16	Optimum mix ratio of high strength mortar	ANOVA method with reread test	$L_9(3^4)$	3-2-3	
Selection of match ratio	17	The loss of collapse degree	Selection of concrete production conditions	$L_4(2^3)$	1-2-2	
	18	The Yanjiu of concrete in this hot area	① Optimization of concrete production conditions ② The 9 test condusion is roughly the some as that of 81 comprehensive test ③Intuitive explanation of the basic principles of orthogonal design	$L_9(3^4)$	1-3-1	

Appendix II Whole Book Case Index

Continued

Category	No.	Project	Characteristic	Orthogonal table	Case number	Page number
Special concrete	19	Development of 800 – 1 000 high strength concrete	①Selection of the best dosage of slag, gypsum and iron powder ②The basic method of equal level orthogonal desyn	$L_9(3^4)$	1-2-1	
			Variance analysis method and basic characteristics of orthogonal design		3-2-1	
	20		The optimum dosage of Mt-150 additive is optimized, and the early strength high-strength concrete with strength greater than 1 000 is selected.	$L_{12}(3^1 \times 2^4)$	I-8	
	21	Research on post addition flow concrete	Optimization of the measurement of water reducing agent	$L_{12}(6^1 \times 2^2)$	I-6	
	22		①Optimization of this condition and mix ratio ②Orthogonal design method with different horizontal numbers (mixed level)	$L_8(4^1 \times 2^4)$	2-1-1	
			A method of variance analysis with repeated sampling		3-2-4	
			Multiple comparison		4-3-2	
	23	No. 600 lightweight aggregate concrete	①Select No. 600 aggregate concrete mix, ②multiple comparisons.	$L_{16}(4^3 \times 2^6)$	I-9	
	24	steel fiber reinforced concrete	The influence of steel fiber content on compressive strength, bending strength and slump is investigated to select the mix proportion of concrete.	$L_8(4^1 \times 2^4)$	I-10	

Continued

Category	No.	Project	Characteristic	Orthogonal table	Case number	Page number
The technology of steam curing and the new technology of	25	Selection of molding and steaming technology	Optimization of admixture and steam curing process, experimental arrangement and analysis method with interaction	$L_8(2^7)$	5-3-1	
			Miscellaneous skills	$L_4(2^3)$	5-4-1	
	26	Quick strength measurement of concrete	Optimization of rapid hardening technology for mortar and investigation of two stage interaction	$L_{16}(2^{15})$	5-3-2	
Construction technology of concrete	27	Wet fly ash	Optimization of technological parameters for wet mixing of fly ash and pseudo level variance analysis method	$L_{18}(6^1 \times 3^5)$	3-2-4	
	28	Production of artificial sand	Influence of artificial sand processing conditions on fineness modulus, variance analysis of orthogonal table with degree of freedom unsaturated	$L_{18}(3^7)$	3-2-6	
Other	29	The effect of the maximum aggregate size on the compressive strength	The analytical method of quasi level	$L_{24}(6^1 \times 4^1 \times 2^3)$	2-2-1	
	30	Determination of water content on the surface of stone	The simplified calculation of variance analysis is used to test the significance of the factors with the stress test error.	$L_{12}(6^1 \times 2^2)$	4-2-3	
			level of operation of the tester.		4-3-3	

Appendix III

Annexed Table

Table III.1 Orthogonal Table

Table III.1.1 $L_4(2^3)$

No.	List number		
	1	2	3
1	1	1	1
2	1	2	2
3	2	1	2
4	2	2	1

Note: The interaction between any two columns is the remaining one

Table III.1.2.1 $L_8(2^7)$

No.	List number						
	1	2	3	4	5	6	7
1	1	2	1	1	1	1	1
2	1	1	1	2	2	2	2
3	1	2	2	1	1	2	2
4	1	2	2	2	2	1	1
5	2	1	2	1	2	1	2
6	2	1	2	2	1	2	1
7	2	2	1	1	2	2	1
8	2	2	1	2	1	1	2

Table III.1.2.2 $L_8(2^7)$ Table of interaction between two columns

No.	List number					
	1	2	3	4	5	6
7	6	5	4	3	2	1
6	7	4	5	2	3	
5	4	7	6	1		
4	5	6	7			
3	3	1				
2	2					

Table III.1.2.3 $L_8(2^7)$ Table of interaction between two columns

No.	List number						
	1	2	3	4	5	6	7
3	A	B	$A \times B$	C	$A \times C$	$B \times C$	
4	A	B	$A \times B$ $C \times D$	C	$A \times C$ $B \times D$	$B \times C$ $A \times D$	D
5	A $D \times E$	B $C \times D$	$A \times B$ $C \times E$	C $B \times D$	$A \times C$ $B \times E$	D $A \times E$ $B \times C$	E $A \times D$

Table III.1.3 $L_{12}(2^{11})$

No.	List number										
	1	2	3	4	5	6	7	8	9	10	11
1	1	1	1	1	1	1	1	1	1	1	1
2	1	1	1	1	1	2	2	2	2	2	2
3	1	1	2	2	2	1	1	1	2	2	2
4	1	2	1	2	2	1	2	2	1	1	2
5	1	2	2	1	2	2	1	2	1	2	1
6	1	2	2	2	1	2	2	1	2	1	1
7	2	1	2	2	1	1	2	2	1	2	1
8	2	1	2	1	2	2	2	1	1	1	2
9	2	1	1	2	2	2	1	2	2	1	1
10	2	2	2	1	1	1	1	2	2	1	2
11	2	2	1	2	1	2	1	1	1	2	2
12	2	2	1	1	2	1	2	1	2	2	1

Appendix III Annexed Table

Table III.1.4.1 $L_{16}(2^{15})$

No.	List number														
	1	2	3	4	5	6	7	8	9	10	11	12	13	14	15
1	1	1	1	1	1	1	1	1	1	1	1	1	1	1	1
2	1	1	1	1	1	1	1	2	2	2	2	2	2	2	2
3	1	1	1	2	2	2	2	1	1	1	1	2	2	2	2
4	1	1	1	2	2	2	2	2	2	2	2	1	1	1	1
5	1	2	2	1	1	2	2	1	1	2	2	1	1	2	2
6	1	2	2	1	1	2	2	2	2	1	1	2	2	1	1
7	1	2	2	2	2	1	1	1	1	2	2	2	2	1	1
8	1	2	2	2	2	1	1	2	2	1	1	1	1	2	2
9	2	1	2	1	1	1	2	1	2	1	2	1	2	1	2
10	2	1	2	1	2	1	2	2	1	2	1	2	1	2	1
11	2	1	2	2	1	2	1	1	2	1	2	2	1	2	1
12	2	1	2	2	1	2	1	2	1	2	1	1	2	1	2
13	2	2	1	1	2	2	1	1	2	2	1	1	2	2	1
14	2	2	1	1	2	2	1	2	1	1	2	2	1	1	2
15	2	2	1	2	1	1	2	1	2	2	1	2	1	1	2
16	2	2	1	2	1	1	2	2	1	1	2	1	2	2	1

Table III.1.4.2 $L_{16}(2^{15})$ Table of interaction between two columns

No.	List number													
	1	2	3	4	5	6	7	8	9	10	11	12	13	14
15	14	13	12	11	10	9	8	7	6	5	4	3	2	1
14	15	12	13	10	11	8	9	6	7	4	5	2	3	
13	12	15	14	9	8	11	10	5	4	7	6	1		
12	13	14	15	8	9	10	11	4	5	6	7			
11	10	9	8	15	14	13	12	3	2	1				
10	11	8	9	14	15	12	13	2	3					
9	8	11	10	13	12	15	14	1						
8	9	10	11	12	13	14	15							
7	6	5	4	3	2	1								
6	7	4	5	2	3									
5	4	7	6	1										
4	5	6	7											
3	2	1												
2	3													

· 155 ·

Table III.1.4.3 $L_{16}(2^{15})$ Head design

No.	List number														
	1	2	3	4	5	6	7	8	9	10	11	12	13	14	15
4	A	B	$A \times B$	C	$A \times C$	$B \times C$		D	$A \times D$	$B \times D$		$C \times D$			E
5	A	B	$A \times B$	C	$A \times C$	$B \times C$	$D \times E$	D	$A \times D$	$B \times D$	$C \times E$	$C \times D$	$B \times E$	$A \times E$	
6	A	B	$A \times B$ $D \times E$	C	$A \times C$ $D \times F$	$B \times C$ $E \times F$		D	$A \times D$ $B \times E$ $C \times F$	$B \times D$ $A \times E$	E	$C \times D$ $A \times F$	F		$C \times E$ $B \times F$
7	A	B	$A \times B$ $D \times E$ $F \times G$	C	$A \times C$ $D \times F$ $E \times G$	$B \times C$ $E \times F$ $D \times G$		D	$A \times D$ $B \times E$ $C \times F$	$B \times D$ $A \times E$ $C \times G$	E	$C \times D$ $A \times F$ $B \times G$	F	G	$C \times E$ $B \times F$ $A \times G$
8	A	B	$A \times B$ $D \times E$ $F \times G$ $C \times H$	C	$A \times C$ $D \times F$ $E \times G$ $B \times H$	$B \times C$ $E \times F$ $D \times G$ $A \times H$	H	D	$A \times D$ $B \times E$ $C \times F$ $G \times H$	$B \times D$ $A \times E$ $C \times G$ $F \times H$	E	$C \times D$ $A \times F$ $B \times G$ $E \times H$	F	G	$C \times E$ $B \times F$ $A \times G$ $D \times H$

Appendix III Annexed Table

Table III.1.5 $L_9(3^4)$

No.	List number			
	1	2	3	4
1	1	1	1	1
2	1	2	2	2
3	1	3	3	3
4	2	1	2	3
5	2	2	3	1
6	2	3	1	2
7	3	1	3	2
8	3	2	1	3
9	3	3	2	1

Table III.1.6.1 $L_{27}(3^{13})$

No.	List number												
	1	2	3	4	5	6	7	8	9	10	11	12	13
1	1	1	3	2	1	2	2	3	1	2	1	3	3
2	2	1	1	1	1	1	3	3	2	1	1	2	1
3	3	1	2	3	1	3	1	3	3	3	1	1	2
4	1	2	2	1	1	2	2	2	3	1	3	1	1
5	2	2	3	3	1	1	3	2	1	3	3	3	2
6	3	2	1	2	1	3	1	2	2	2	3	2	3
7	1	3	1	3	1	2	2	1	2	3	2	2	2
8	2	3	2	2	1	1	3	1	3	2	2	1	3
9	3	3	3	1	1	3	1	1	1	1	2	3	1
10	1	1	1	1	2	3	3	1	3	2	3	3	2
11	2	1	2	3	2	2	1	1	1	1	3	2	3
12	3	1	3	2	2	1	2	1	2	3	3	1	1
13	1	2	3	3	2	3	3	3	2	1	2	1	3
14	2	2	1	2	2	2	1	3	3	3	2	1	2
15	3	2	2	1	2	1	2	3	1	2	2	2	2
16	1	3	2	2	2	3	3	2	1	3	1	2	1
17	2	3	3	1	2	2	1	2	2	2	1	1	2
18	3	3	1	3	2	1	2	2	3	1	1	3	3
19	1	1	2	3	3	1	1	2	2	2	2	3	1

Continued Table III.1.6.1

No.	List number												
	1	2	3	4	5	6	7	8	9	10	11	12	13
20	2	1	3	2	3	3	2	2	3	1	2	2	2
21	3	1	1	1	3	2	3	2	1	3	2	1	3
22	1	2	1	2	3	1	1	1	1	1	1	1	2
23	2	2	2	1	3	3	2	1	2	3	1	3	3
24	3	2	3	3	3	2	3	1	3	2	1	2	1
25	1	3	3	1	3	1	1	3	3	3	3	2	3
26	2	3	1	3	3	3	2	3	1	2	3	1	1
27	3	3	2	2	3	2	3	3	2	1	3	3	2

Table III.1.6.2 $L_{27}(3^{13})$ Table of interaction between two columns

No.	List number											
	1	2	3	4	5	6	7	8	9	10	11	12
13	11 12	7 10	5 9	6 8	3 9	4 8	2 10	4 6	3 5	2 7	1 12	1 11
12	11 13	6 9	7 8	5 10	4 10	2 9	3 8	3 7	2 6	4 5	1 13	
11	12 13	5 8	6 10	7 9	2 8	3 10	4 9	2 5	4 7	3 6		
10	8 9	7 13	6 11	5 12	4 12	3 11	2 13	1 9	1 8			
9	8 10	6 12	5 13	7 11	3 13	2 12	4 11	1 10				
8	9 10	5 11	7 12	6 13	2 11	4 13	3 12					
7	5 6	10 13	8 12	9 11	1 6	1 5						
6	5 7	9 12	10 11	8 13	1 7							
5	6 7	8 11	9 13	10 12								
4	2 3	1 3	1 2									
3	2 4	1 4										
2	3 4											

Appendix III Annexed Table

Table III.1.6.3 $L_{27}(3^{13})$ Table of interaction between two columns

No.	List number						
	1	2	3	4	5	6	7
3	A	B	$(A \times B)_1$	$(A \times B)_2$	C	$(A \times C)_1$	$(A \times C)_2$
4	A	B	$(A \times B)_1$ $(C \times D)_2$	$(A \times B)_2$	C	$(A \times C)_1$ $(B \times D)_2$	$(A \times C)_2$

No.	List number					
	8	9	10	11	12	13
3	$(B \times C)_1$			$(B \times C)_2$		
4	$(B \times C)_1$ $(A \times D)_2$	D	$(A \times B)_1$	$(B \times C)_2$	$(B \times D)_1$	$(C \times D)_1$

Table III.1.7 $L_{16}(4^5)$

No.	List number				
	1	2	3	4	5
1	1	2	3	2	3
2	3	4	1	2	2
3	2	4	3	3	4
4	4	2	1	3	1
5	1	3	1	4	4
6	3	1	3	4	1
7	2	1	1	1	3
8	4	3	3	1	2
9	1	1	4	3	2
10	3	3	2	3	3
11	2	3	4	2	1
12	4	1	2	2	4
13	1	4	2	1	1
14	3	2	4	1	4
15	2	2	2	4	2
16	4	4	4	4	3

Note: Any two columus snferact with the other fourcolumns

Table III.1.8 $L_{25}(5^6)$

No.	List number					
	1	2	3	4	5	6
1	1	1	2	4	3	2
2	2	1	5	5	5	4
3	3	1	4	1	4	1
4	4	1	1	3	1	3
5	5	1	3	2	2	5
6	1	2	3	3	4	4
7	2	2	2	2	1	1
8	3	2	5	4	2	3
9	4	2	4	5	3	5
10	5	2	1	1	5	2
11	1	3	1	5	2	1
12	2	3	3	1	3	3
13	3	3	2	3	5	5
14	4	3	5	2	4	2
15	5	3	4	4	1	4
16	1	4	4	2	5	3
17	2	4	1	4	4	5
18	3	4	3	5	1	2
19	4	4	2	1	2	4
20	5	4	5	3	3	1
21	1	5	5	1	1	5
22	2	5	4	3	2	2
23	3	5	1	2	3	4
24	4	5	3	4	4	1
25	5	5	2	5	5	3

Table III.1.9 $L_8(4^1 \times 2^4)$

No.	List number				
	1	2	3	4	5
1	1	1	2	2	1
2	3	2	2	1	1
3	2	2	2	2	2

Continued Table III.1.9

No.	List number				
	1	2	3	4	5
4	4	1	2	1	2
5	1	2	1	1	2
6	3	1	1	2	2
7	2	1	1	1	1
8	4	2	1	2	1

Table III.1.10 $L_{12}(3^1 \times 2^4)$

No.	List number				
	1	2	3	4	5
1	2	1	1	1	2
2	2	2	1	2	1
3	2	1	2	2	2
4	2	2	2	1	1
5	1	1	1	2	2
6	1	2	1	2	1
7	1	1	2	1	1
8	1	2	2	1	2
9	3	1	1	1	1
10	3	2	1	1	2
11	3	1	2	2	1
12	3	2	2	2	2

Table III.1.11 $L_{16}(4^4 \times 2^3)$

No.	List number						
	1	2	3	4	5	6	7
1	1	2	3	2	2	1	2
2	3	4	1	2	1	2	2
3	2	4	3	3	2	2	1
4	4	2	1	3	1	1	1
5	1	3	1	4	2	2	1
6	3	1	3	4	1	1	1
7	2	1	1	1	2	1	2

Continued Table III.1.11

No.	List number						
	1	2	3	4	5	6	7
8	4	3	3	1	1	2	2
9	1	1	4	3	1	2	2
10	3	3	2	3	2	1	2
11	2	3	4	2	1	1	1
12	4	1	2	2	2	2	1
13	1	4	2	1	1	1	1
14	3	2	4	1	2	2	1
15	2	2	2	4	1	2	2
16	4	4	4	4	2	1	2

Table III.1.12 $L_{16}(4^3 \times 2^6)$

No.	List number								
	1	2	3	4	5	6	7	8	9
1	1	2	3	1	2	2	1	1	2
2	3	4	1	1	1	2	2	1	2
3	2	4	3	2	2	1	2	1	1
4	4	2	1	2	1	1	1	1	1
5	1	3	1	2	2	2	2	2	1
6	3	1	3	2	1	2	1	2	1
7	2	1	1	1	2	1	1	2	2
8	4	3	3	1	1	1	2	2	2
9	1	1	4	2	1	1	2	1	2
10	3	3	2	2	2	1	1	1	2
11	2	3	4	1	1	2	1	1	1
12	4	1	2	1	2	2	2	1	1
13	1	4	2	1	1	1	1	2	1
14	3	2	4	1	2	1	2	2	1
15	2	2	2	2	1	2	2	2	2
16	4	4	4	2	2	2	1	2	2

Appendix III Annexed Table

Table III.1.13 $L_{16}(4^2 \times 2^9)$

No.	List number										
	1	2	3	4	5	6	7	8	9	10	11
1	1	2	2	1	1	2	2	1	1	1	2
2	3	4	1	1	1	1	2	2	1	2	2
3	2	4	2	2	1	2	1	2	1	1	1
4	4	2	1	2	1	1	1	1	1	2	1
5	1	3	1	2	1	2	2	2	2	2	1
6	3	1	2	2	1	1	2	1	2	1	1
7	2	1	1	1	1	2	1	1	2	2	2
8	4	3	2	1	1	1	1	2	2	1	2
9	1	1	2	2	2	1	1	2	1	2	2
10	3	3	1	2	2	2	1	1	1	1	2
11	2	3	2	1	2	1	2	1	1	2	1
12	4	1	1	1	2	2	2	2	1	1	1
13	1	4	1	1	2	1	1	1	2	1	1
14	3	2	2	1	2	2	1	2	2	2	1
15	2	2	1	2	2	1	2	2	2	1	2
16	4	4	2	2	2	2	2	1	2	2	2

Table III.1.14.1 $L_{16}(4 \times 2^{12})$

No.	List number												
	1	2	3	4	5	6	7	8	9	10	11	12	13
1	1	1	2	2	1	2	1	2	2	1	1	1	2
2	3	2	2	1	1	1	1	1	2	2	1	2	2
3	2	2	2	2	2	1	1	2	1	2	1	1	1
4	4	1	2	1	2	2	1	1	1	1	1	2	1
5	1	2	1	1	2	2	1	2	2	2	2	2	1
6	3	1	1	2	2	1	1	1	2	1	2	1	1
7	2	1	1	1	1	1	1	2	1	1	2	2	2
8	4	2	1	2	1	2	1	1	1	2	2	1	2
9	1	1	1	2	2	1	2	1	1	2	1	2	2
10	3	2	1	1	2	2	2	2	1	1	1	1	2
11	2	2	1	2	1	2	2	1	2	1	1	2	1
12	4	1	1	1	1	1	2	2	2	2	1	1	1

Continued Table III.1.14.1

No.	List number												
	1	2	3	4	5	6	7	8	9	10	11	12	13
13	1	2	2	1	1	1	2	1	1	1	2	1	1
14	3	1	2	2	1	2	2	2	1	2	2	2	1
15	2	1	2	1	2	2	2	1	2	2	2	1	2
16	4	2	2	2	2	1	2	2	2	1	2	2	2

Table III.1.14.2 $L_{16}(4 \times 2^{12})$ Table of interaction between two columns

No.	List number						
	1	2	3	4	5	6	7
3	A	B	$(A \times B)_1$	$(A \times B)_2$	$(A \times B)_3$	C	$(A \times C)_1$
4	A	B	$(A \times B)_1$ $C \times D$	$(A \times B)_2$	$(A \times B)_3$	C	$(A \times C)_1$ $B \times D$
5	A	B	$(A \times B)_1$ $C \times D$	$(A \times B)_2$ $C \times E$	$(A \times B)_3$	C	$(A \times C)_1$ $B \times D$

No.	List number					
	8	9	10	11	12	13
3	$(A \times C)_2$	$(A \times C)_3$	$B \times C$			
4	$(A \times C)_2$	$(A \times C)_3$	$B \times C$ $(A \times D)_1$	D	$(A \times D)_3$	$(A \times D)_2$
5	$(A \times C)_2$ $B \times E$	$(A \times C)_3$	$B \times C$ $(A \times D)_1$ $(A \times E)_2$	D $(A \times E)_3$	E $(A \times D)_3$	$(A \times E)_1$ $(A \times D)_2$

Table III.1.15 $L_8(2^1 \times 3^7)$

No.	List number							
	1	2	3	4	5	6	7	8
1	1	1	1	3	2	2	1	2
2	1	2	1	1	1	1	2	1
3	1	3	1	2	3	3	3	3
4	1	1	2	2	1	2	3	1
5	1	2	2	3	3	1	1	3
6	1	3	2	1	2	3	2	2
7	1	1	3	1	3	1	3	2
8	1	2	3	2	2	3	1	1

Continued Table Ⅲ.1.15

No.	List number							
	1	2	3	4	5	6	7	8
9	1	3	3	3	1	2	2	3
10	2	1	1	1	1	3	1	3
11	2	2	1	2	3	2	2	2
12	2	3	1	3	2	1	3	1
13	2	1	2	3	3	3	2	1
14	2	2	2	1	2	2	3	3
15	2	3	2	2	1	1	1	2
16	2	1	3	2	2	1	2	3
17	2	2	3	3	1	3	3	2
18	2	3	3	1	3	2	1	1

Table Ⅲ.1.16 $L_{13}(6 \times 3^6)$

No.	List number						
	1	2	3	4	5	6	7
1	1	1	3	2	2	1	2
2	1	2	1	1	1	2	1
3	1	3	2	3	3	3	3
4	2	1	2	1	2	3	1
5	2	2	3	3	1	1	3
6	2	3	1	2	3	2	2
7	3	1	1	3	1	3	2
8	3	2	2	2	3	1	1
9	3	3	3	1	2	2	3
10	4	1	1	1	3	1	3
11	4	2	2	3	2	2	2
12	4	3	3	2	1	3	1
13	5	1	3	3	3	2	1
14	5	2	1	2	2	3	3
15	5	3	2	1	1	1	2
16	6	1	2	2	1	2	3
17	6	2	3	1	3	3	2
18	6	3	1	3	2	1	1

Table III.1.17 $L_{12}(6 \times 2^2)$

No.	List number		
	1	2	3
1	1	1	1
2	2	1	2
3	1	2	2
4	2	2	1
5	3	1	2
6	4	1	1
7	3	2	1
8	4	2	2
9	5	1	1
10	6	1	2
11	5	2	2
12	6	2	1

Table III.1.18 $L_{20}(5 \times 2^8)$

No.	List number								
	1	2	3	4	5	6	7	8	9
1	1	1	1	1	1	1	1	1	1
2	1	1	1	1	1	2	2	2	2
3	1	2	2	2	2	1	1	1	1
4	1	2	2	2	2	2	2	2	2
5	2	1	2	1	2	1	1	1	2
6	2	1	2	2	1	1	2	2	1
7	2	2	1	1	2	2	1	2	1
8	2	2	1	2	1	2	2	1	2
9	3	1	1	2	1	1	1	2	2
10	3	1	2	2	2	2	2	1	1
11	3	2	1	1	2	1	2	2	1
12	3	2	2	1	1	2	1	1	2
13	4	1	1	2	2	1	2	1	2
14	4	1	2	1	2	2	1	2	2
15	4	2	1	2	1	2	1	1	1
16	4	2	2	1	1	1	2	2	1

Continued Table III.1.18

No.	List number								
	1	2	3	4	5	6	7	8	9
17	5	1	1	1	2	2	2	1	1
18	5	1	2	2	1	2	1	2	1
19	5	2	1	2	2	1	1	2	2
20	5	2	2	1	1	1	2	1	2

Table III.1.19 $L_{24}(3^1 \times 4^1 \times 2^4)$

No.	List number					
	1	2	3	4	5	6
1	1	1	1	1	1	1
2	1	2	1	1	2	2
3	1	3	1	2	2	1
4	1	4	1	2	1	2
5	1	1	2	2	2	2
6	1	2	2	2	1	1
7	1	3	2	1	1	2
8	1	4	2	1	2	1
9	2	1	1	1	1	2
10	2	2	1	1	2	1
11	2	3	1	2	2	2
12	2	4	1	2	1	1
13	2	1	2	2	2	1
14	2	2	2	2	1	2
15	2	3	2	1	1	1
16	2	4	2	1	2	2
17	3	1	1	1	1	2
18	3	2	1	1	2	1
19	3	3	1	2	2	2
20	3	4	1	2	1	1
21	3	1	2	2	2	1
22	3	2	2	2	1	2
23	3	3	2	1	1	1
24	3	4	2	1	2	2

Orthogonal Design in Concrete Application

Table III.1.20 $L_{24}(6^1 \times 4^1 \times 2^3)$

No.	List number				
	1	2	3	4	5
1	1	1	1	1	2
2	1	2	1	2	1
3	1	3	2	2	2
4	1	4	2	1	1
5	2	1	2	2	1
6	2	2	2	1	2
7	2	3	1	1	1
8	2	4	1	2	2
9	3	1	1	1	1
10	3	2	1	2	2
11	3	3	2	2	1
12	3	4	2	1	2
13	4	1	2	2	2
14	4	2	2	1	1
15	4	3	1	1	2
16	4	4	1	2	1
17	5	1	1	1	1
18	5	2	1	2	2
19	5	3	2	2	1
20	5	4	2	1	2
21	6	1	2	2	2
22	6	2	2	1	1
23	6	3	1	1	2
24	6	4	1	2	1

Table III.1.21 $L_{35}(4^1 \times 3^{10})$

No.	List number										
	1	2	3	4	5	6	7	8	9	10	11
1	1	1	1	3	2	2	1	2	1	3	3
2	1	2	1	1	1	1	2	1	3	2	1
3	1	3	1	2	3	3	3	3	2	1	2
4	1	1	2	2	1	2	3	1	1	3	3

Continued Table III.1.21

No.	List number										
	1	2	3	4	5	6	7	8	9	10	11
5	1	2	2	3	3	1	1	3	3	2	1
6	1	3	2	1	2	3	2	2	2	1	2
7	1	1	3	1	3	1	3	2	1	2	2
8	1	2	3	2	2	3	1	1	3	1	3
9	1	3	3	3	1	2	2	3	2	3	1
10	3	1	1	1	1	3	1	3	3	3	2
11	3	2	1	2	3	2	2	2	2	2	3
12	3	3	1	3	2	1	3	1	1	1	1
13	3	1	2	3	3	3	2	1	3	3	2
14	3	2	2	1	2	2	3	3	2	2	3
15	3	3	2	2	1	1	1	2	1	1	1
16	3	1	3	2	2	1	2	3	1	2	2
17	3	2	3	3	1	3	3	2	3	1	3
18	3	3	3	1	3	2	1	1	2	3	1
19	4	1	1	2	3	2	3	3	3	1	1
20	4	2	1	3	2	1	1	2	2	3	2
21	4	3	1	1	1	3	2	1	1	2	3
22	4	1	2	1	2	2	2	2	3	1	1
23	4	2	2	2	1	1	3	1	2	3	2
24	4	3	2	3	3	3	1	3	1	2	3
25	4	1	3	3	1	1	2	3	2	1	3
26	4	2	3	1	3	3	3	2	1	3	1
27	4	3	3	2	2	2	1	1	3	2	2
28	2	1	1	3	2	3	3	1	2	2	1
29	2	2	1	1	1	2	1	3	1	1	2
30	2	3	1	2	2	1	2	2	3	3	3
31	2	1	2	2	1	3	1	2	2	2	1
32	2	2	2	3	3	2	2	1	1	1	2
33	2	3	2	1	2	1	3	3	3	3	3
34	2	1	3	1	3	1	1	1	2	1	3
35	2	2	3	2	2	3	2	3	1	3	1
36	2	3	3	3	1	2	3	2	3	2	2

Orthogonal Design in Concrete Application

Table III.1.22 $L_{36}(2^1 \times 6^2 \times 3^5)$

No.	List number							
	1	2	3	4	5	6	7	8
1	1	1	1	3	2	2	1	2
2	1	2	1	1	1	1	2	1
3	1	3	1	2	3	3	3	3
4	1	1	2	2	1	2	3	1
5	1	2	2	3	3	1	1	3
6	1	3	2	1	2	3	2	2
7	1	1	3	1	3	1	3	2
8	1	2	3	2	2	3	1	1
9	1	3	3	3	1	2	2	3
10	2	1	4	1	1	3	1	3
11	2	2	4	2	3	2	2	2
12	2	3	4	3	2	1	3	1
13	2	1	5	3	3	3	2	1
14	2	2	5	1	2	2	3	3
15	2	3	5	2	1	1	1	2
16	2	1	6	2	2	1	2	3
17	2	2	6	3	1	3	3	2
18	2	3	6	1	3	2	1	1
19	2	4	1	2	3	2	3	3
20	2	5	1	3	2	1	1	2
21	2	6	1	1	1	3	2	1
22	2	4	2	1	2	2	2	2
23	2	5	2	2	1	1	3	1
24	2	6	2	3	3	3	1	3
25	2	4	3	3	1	1	2	3
26	2	5	3	1	3	3	3	2
27	2	6	3	2	2	2	1	1
28	1	4	4	3	2	3	3	1
29	1	5	4	1	1	2	1	3
30	1	6	4	2	3	1	2	2
31	1	4	5	2	1	3	1	2

Continued Table III.1.22

No.	List number							
	1	2	3	4	5	6	7	8
32	1	5	5	3	3	2	2	1
33	1	6	5	1	2	1	3	3
34	1	4	6	1	3	1	1	1
35	1	5	6	2	2	3	2	3
36	1	6	6	3	1	2	3	2

Table III.1.23 $L_{36}(6^3 \times 3^3)$

No.	List number					
	1	2	3	4	5	6
1	1	1	3	2	1	2
2	1	2	1	1	2	1
3	1	3	2	3	3	3
4	2	1	2	1	3	1
5	2	2	3	3	1	3
6	2	3	1	2	2	2
7	3	1	1	3	3	2
8	3	2	2	2	1	1
9	3	3	3	1	2	3
10	4	1	4	1	1	3
11	4	2	5	3	2	2
12	4	3	6	2	3	1
13	5	1	6	3	2	1
14	5	2	4	2	3	3
15	5	3	5	1	1	2
16	6	1	5	2	2	3
17	6	2	6	1	3	2
18	6	3	4	3	1	1
19	1	4	5	3	3	3
20	1	5	6	2	1	2
21	1	6	4	1	2	1
22	2	4	4	2	2	2
23	2	5	5	1	3	1

Continued Table Ⅲ.1.23

No.	List number					
	1	2	3	4	5	6
24	2	6	6	3	1	3
25	3	4	6	1	2	3
26	3	5	4	3	3	2
27	3	6	5	2	1	1
28	4	4	3	2	3	1
29	4	5	1	1	1	3
30	4	6	2	3	2	2
31	5	4	2	1	1	2
32	5	5	3	3	2	1
33	5	6	1	2	3	3
34	6	4	1	3	1	1
35	6	5	2	2	2	3
36	6	6	3	1	3	2

Table Ⅲ.1.24 $L_{15}(8^1 \times 2^8)$

No.	List number								
	1	2	3	4	5	6	7	8	9
1	1	2	1	2	1	2	2	1	1
2	2	2	1	1	1	1	2	2	2
3	3	2	2	1	1	2	1	2	1
4	4	2	2	2	1	1	1	1	2
5	5	1	2	2	1	2	2	2	2
6	6	1	2	1	1	1	2	1	1
7	7	1	1	1	1	2	1	1	2
8	8	1	1	2	1	1	1	2	1
9	1	1	2	1	2	1	1	2	2
10	2	1	2	2	2	2	1	1	1
11	3	1	1	2	2	1	2	1	2
12	4	1	1	1	2	2	2	2	1
13	5	2	1	1	2	1	1	1	1
14	6	2	1	2	2	2	1	2	2
15	7	2	2	2	2	1	2	2	1
16	8	2	2	1	2	2	2	1	2

Appendix III Annexed Table

Table III.2.1 $F(a=0.20)$

f_2 \ f_1	1	2	3	4	5	6	12	24	∞
1	9.5	12.0	13.1	13.7	14.0	14.3	14.9	15.2	51.6
2	3.6	4.0	4.2	4.2	4.3	4.3	4.4	4.4	4.5
3	2.7	2.9	2.9	3.0	3.0	3.0	3.0	3.0	3.0
4	2.4	2.5	2.5	2.5	2.5	2.5	2.5	2.4	2.4
5	2.2	2.3	2.3	2.2	2.2	2.2	2.2	2.2	2.1
6	2.1	2.1	2.1	2.1	2.1	2.1	2.0	2.0	2.0
7	2.0	2.0	2.0	2.0	2.0	2.0	1.9	1.9	1.8
8	2.0	2.0	1.9	1.9	1.9	1.9	1.8	1.8	1.7
9	1.9	1.9	1.9	1.9	1.9	1.8	1.7	1.7	1.7
10	1.9	1.9	1.9	1.8	1.8	1.8	1.7	1.7	1.6
11	1.8	1.9	1.8	1.8	1.0	1.8	1.7	1.6	1.6
12	1.8	1.8	1.8	1.8	1.8	1.7	1.7	1.6	1.6
13	1.8	1.8	1.8	1.8	1.7	1.7	1.6	1.6	1.5
14	1.8	1.8	1.7	1.7	1.7	1.7	1.6	1.6	1.5
15	1.8	1.8	1.7	1.7	1.7	1.7	1.6	1.5	1.5
16	1.8	1.8	1.7	1.7	1.7	1.6	1.6	1.5	1.5
17	1.8	1.8	1.7	1.7	1.7	1.6	1.6	1.5	1.4
18	1.8	1.8	1.7	1.7	1.6	1.6	1.5	1.5	1.4
19	1.8	1.8	1.7	1.7	1.6	1.6	1.5	1.5	1.4
20	1.8	1.8	1.7	1.7	1.6	1.6	1.5	1.5	1.4
22	1.8	1.7	1.7	1.6	1.6	1.6	1.5	1.4	1.4
24	1.7	1.7	1.6	1.6	1.6	1.6	1.5	1.4	1.4
26	1.7	1.7	1.6	1.6	1.6	1.6	1.5	1.4	1.3
28	1.7	1.7	1.7	1.6	1.6	1.6	1.5	1.4	1.3
30	1.7	1.7	1.6	1.6	1.6	1.5	1.5	1.4	1.3
40	1.7	1.7	1.6	1.6	1.5	1.5	1.4	1.4	1.2
60	1.7	1.7	1.6	1.6	1.5	1.5	1.4	1.3	1.2
120	1.6	1.6	1.5	1.5	1.5	1.5	1.4	1.3	1.1
∞	1.6	1.6	1.5	1.5	1.5	1.5	1.3	1.2	1.0

Table Ⅲ.2.2　$F(a=0.1)$

f_2 \ f_1	1	2	3	4	5	6	12	24	∞
1	39.9	49.5	53.6	55.8	57.2	58.2	60.7	62.0	63.3
2	8.5	9.0	9.2	9.2	9.3	9.3	9.4	9.4	9.5
3	5.5	5.5	5.4	5.3	5.3	5.3	5.2	5.2	5.1
4	4.5	4.3	4.2	4.1	4.1	4.0	3.9	3.8	3.8
5	4.1	3.8	3.6	3.5	3.5	3.4	3.3	3.2	3.1
6	3.8	3.5	3.3	3.2	3.1	3.1	2.9	2.8	2.7
7	3.6	3.3	3.1	3.0	2.9	2.8	2.7	2.6	2.5
8	3.5	3.1	2.9	2.8	2.7	2.7	2.5	2.4	2.3
9	3.4	3.0	2.8	2.7	2.6	2.6	2.4	2.3	2.2
10	3.3	2.9	2.7	2.6	2.5	2.5	2.3	2.2	2.1
11	3.2	2.9	2.7	2.5	2.5	2.4	2.2	2.1	2.0
12	3.2	2.8	2.6	2.5	2.4	2.3	2.1	2.0	1.9
13	3.1	2.8	2.4	2.4	2.3	2.3	2.1	2.0	1.8
14	3.1	2.7	2.4	2.4	2.3	2.2	2.1	1.9	1.8
15	3.1	2.7	2.4	2.4	2.3	2.2	2.0	1.9	1.8
16	3.0	2.7	2.4	2.3	2.2	2.2	2.0	1.9	1.7
17	3.0	2.7	2.3	2.3	2.2	2.2	2.0	1.8	1.7
18	3.0	2.6	2.4	2.3	2.2	2.1	1.9	1.8	1.7
19	3.0	2.6	2.4	2.3	2.2	2.1	1.9	1.8	1.6
20	3.0	2.6	2.4	2.2	2.2	2.1	1.9	1.8	1.6
22	2.9	2.6	2.4	2.2	2.1	2.1	1.9	1.7	1.6
24	2.9	2.6	2.3	2.2	2.1	2.0	1.8	1.7	1.5
26	2.9	2.6	2.3	2.2	2.1	2.0	1.8	1.7	1.5
28	2.9	2.5	2.3	2.2	2.1	2.0	1.8	1.7	1.5
30	2.9	2.5	2.3	2.1	2.0	2.0	1.8	1.6	1.5
40	2.8	2.4	2.2	2.1	2.0	1.9	1.7	1.6	1.4
60	2.8	2.4	2.2	2.0	1.9	1.9	1.7	1.5	1.3
120	2.7	2.3	2.1	2.0	1.9	1.8	1.6	1.4	1.2
∞	2.7	2.3	2.1	1.9	1.8	1.8	1.5	1.4	1.0

Appendix III Annexed Table

Table III.2.3 $F(a=0.05)$

f_2 \ f_1	1	2	3	4	5	6	12	24	∞
1	161.4	199.5	215.7	224.6	230.2	234.0	234.9	249.0	254.3
2	18.5	19.0	19.2	19.3	19.3	19.3	19.4	19.5	19.5
3	10.1	9.6	9.3	9.1	9.0	8.9	8.7	8.6	8.5
4	7.7	6.9	6.6	6.4	6.3	6.2	5.9	5.8	5.6
5	6.6	5.8	5.4	5.2	5.1	5.0	4.7	4.5	4.4
6	6.0	5.1	4.8	4.5	4.4	4.3	4.0	3.8	3.7
7	5.6	4.7	4.4	4.1	4.0	3.9	3.6	3.4	3.2
8	5.3	4.5	4.1	3.8	3.7	3.6	3.3	3.1	2.9
9	5.1	4.3	3.9	3.6	3.5	3.4	3.1	2.9	2.7
10	5.0	4.1	3.7	3.5	3.3	3.2	2.9	2.7	2.5
11	4.8	4.0	3.6	3.4	3.2	3.1	2.8	2.6	2.4
12	4.8	3.9	3.5	3.3	3.1	3.0	2.7	2.5	2.3
13	4.7	3.8	3.4	3.2	3.0	2.9	2.6	2.4	2.2
14	4.6	3.7	3.3	3.1	3.0	2.9	2.5	2.3	2.1
15	4.5	3.7	3.3	3.1	2.9	2.8	2.5	2.3	2.1
16	4.5	3.6	3.2	3.0	2.9	2.7	2.4	2.2	2.0
17	4.5	3.6	3.2	3.0	2.8	2.7	2.4	2.2	2.0
18	4.4	3.6	3.2	2.9	2.8	2.7	2.3	2.1	1.9
19	4.4	3.5	3.1	2.9	2.7	2.6	2.3	2.1	1.9
20	4.4	3.5	3.1	2.9	2.7	2.6	2.3	2.1	1.8
22	4.3	3.4	3.1	2.8	2.7	2.6	2.2	2.0	1.8
24	4.3	3.4	3.0	2.8	2.6	2.5	2.2	2.0	1.7
26	4.2	3.4	3.0	2.7	2.6	2.5	2.2	2.0	1.7
28	4.2	3.3	3.0	2.7	2.6	2.4	2.1	1.9	1.7
30	4.2	3.3	2.9	2.7	2.5	2.4	2.1	1.9	1.6
40	4.1	3.2	2.8	2.6	2.5	2.3	2.0	1.8	1.5
60	4.0	3.2	2.8	2.5	2.4	2.3	1.9	1.7	1.4
120	3.9	3.1	2.7	2.5	2.3	2.2	1.8	1.6	1.3
∞	3.8	3.0	2.6	2.4	2.2	2.1	1.8	1.5	1.0

Table III.2.4 $F(a=0.01)$

f_2 \ f_1	1	2	3	4	5	6	12	24	∞
1	4 052	4 999	5 403	5 625	5 764	5 859	6 106	6 234	6 366
2	98.5	99.0	99.2	99.3	99.3	99.3	99.4	99.5	99.5
3	34.1	30.8	29.5	28.7	28.2	27.9	27.1	26.6	26.1
4	21.2	18.0	16.7	16.0	15.5	15.2	14.4	13.9	13.5
5	16.3	13.3	12.1	11.4	11.0	10.7	9.9	9.5	9.0
6	13.7	10.9	9.8	9.2	8.8	8.5	7.7	7.3	6.9
7	12.3	9.6	8.5	7.9	7.5	7.2	6.5	6.1	5.7
8	11.3	8.7	7.6	7.0	6.6	6.4	5.7	5.3	4.9
9	10.6	8.0	7.0	6.4	6.1	5.8	5.1	4.7	4.3
10	10.0	7.6	6.6	6.0	5.6	5.4	4.7	4.3	3.9
11	9.7	7.2	6.2	5.7	5.3	5.1	4.4	4.0	3.6
12	9.3	6.9	6.0	5.4	5.1	4.8	4.2	3.8	3.4
13	9.1	6.7	5.7	5.2	4.9	4.6	4.0	3.6	3.2
14	8.9	6.5	5.6	5.0	4.7	4.5	3.8	3.4	3.0
15	8.7	6.4	5.4	4.9	4.6	4.3	3.7	3.3	2.9
16	8.5	6.2	5.3	4.8	4.4	4.2	3.6	3.2	2.8
17	8.4	6.1	5.2	4.7	4.3	4.1	3.5	3.1	2.7
18	8.3	6.0	5.1	4.6	4.3	4.0	3.4	3.0	2.6
19	8.2	5.9	5.0	4.5	4.2	3.9	3.3	2.9	2.5
20	8.1	5.9	4.9	4.4	4.1	3.9	3.2	2.9	2.4
22	7.9	5.7	4.8	4.3	4.0	3.8	3.1	2.8	2.3
24	7.8	5.6	4.7	4.2	3.9	3.7	3.0	2.7	2.2
26	7.7	5.5	4.6	4.1	3.8	3.6	3.0	2.6	2.1
28	7.6	5.5	4.6	4.1	3.8	3.5	2.9	2.5	2.1
30	7.6	5.4	4.5	4.0	3.7	3.5	2.8	2.5	2.0
40	7.3	5.2	4.3	3.8	3.5	3.3	2.7	2.3	1.8
60	7.1	5.0	4.1	3.7	3.3	3.1	2.5	2.1	1.6
120	6.9	4.8	4.0	3.5	3.2	3.0	2.3	2.0	1.4
∞	6.6	4.6	3.8	3.3	3.0	2.8	2.2	1.8	1.0

Appendix III Annexed Table

Table III.3.1 $q(a=0.10)$

ϕ \ m	2	3	4	5	6	7	8	9	10	15	20
1	8.93	13.40	16.40	18.50	20.20	21.50	22.60	23.60	24.50	27.60	29.70
2	4.13	5.73	6.77	7.54	8.14	8.63	9.05	9.41	9.72	10.90	11.70
3	3.33	4.47	5.20	5.74	6.16	6.51	6.81	7.06	7.29	8.12	8.68
4	3.01	3.98	4.59	5.03	5.39	5.68	5.93	6.14	6.33	7.02	7.50
5	2.85	3.72	4.26	4.66	4.98	5.24	5.46	5.65	5.82	6.44	6.86
6	2.75	3.56	4.07	4.44	4.73	4.97	5.17	5.34	5.50	6.07	6.47
7	2.68	3.45	3.93	4.28	4.55	4.78	4.97	5.14	5.28	5.83	6.19
8	2.63	3.37	3.83	4.17	4.43	4.65	4.83	4.99	5.13	5.64	6.00
9	2.59	3.32	3.76	4.08	4.34	4.54	4.72	4.87	5.01	5.51	5.85
10	2.56	3.27	3.70	4.02	4.26	4.47	4.64	4.78	4.91	5.40	5.739
11	2.54	3.23	3.66	3.96	4.20	4.40	4.57	4.71	4.84	5.31	5.63
12	2.52	3.20	3.62	3.92	4.16	4.35	4.51	4.65	4.78	5.24	5.55
13	2.50	3.18	3.59	3.38	4.12	4.30	4.46	4.60	4.72	5.18	5.48
14	2.49	3.16	3.56	3.85	4.08	4.27	4.42	4.56	4.68	5.12	5.43
15	2.48	3.14	3.54	3.83	4.05	4.23	4.39	4.52	4.64	5.08	5.38
16	2.47	3.12	3.52	3.80	4.03	4.21	4.36	4.49	4.61	5.04	5.33
17	2.46	3.11	3.50	3.78	4.00	4.18	4.33	4.46	4.58	5.01	5.30
18	2.45	3.10	3.49	3.77	3.98	4.16	4.31	4.44	4.55	4.998	5.26
19	2.45	3.09	3.47	3.75	3.97	4.14	4.29	4.42	4.53	4.95	5.23
20	2.44	3.08	3.46	3.74	3.95	4.12	4.27	4.40	4.51	4.92	5.20
24	2.42	3.05	3.42	3.69	3.90	4.07	4.21	4.34	4.44	4.85	5.12
30	2.40	3.02	3.39	3.65	3.85	4.02	4.16	4.28	4.38	4.77	5.03
40	2.38	2.99	3.35	3.60	3.80	3.96	4.10	4.21	4.32	4.69	4.95
60	2.36	2.96	3.31	3.56	3.75	3.91	4.04	4.10	4.25	4.62	4.86
120	2.34	2.93	3.28	3.52	3.71	3.86	3.99	4.10	4.19	4.54	4.78
∞	2.33	2.90	3.24	3.48	3.66	3.81	3.93	4.04	4.18	4.47	4.69

Table III.3.2 $q(a=0.05)$

ϕ \ m	2	3	4	5	6	7	8	9	10	15	20
1	18.0	27.0	32.8	37.1	40.4	43.1	45.4	47.4	49.1	55.4	59.6
2	6.08	8.33	9.80	10.9	11.7	12.4	13.0	13.5	14.0	15.7	16.8
3	4.50	5.91	6.82	7.50	8.04	8.48	8.85	9.18	9.46	10.5	11.2
4	3.93	5.04	5.76	6.29	6.71	7.05	7.35	7.60	7.83	8.66	9.23
5	3.64	4.60	5.22	5.67	6.03	6.33	6.58	6.80	6.99	7.72	8.21
6	3.46	4.34	4.90	5.30	5.63	5.90	6.12	6.32	6.49	7.14	7.59
7	3.34	4.16	4.68	5.06	5.36	5.61	5.82	6.00	6.16	6.76	7.17
8	3.26	4.04	4.53	4.89	5.17	5.40	5.60	5.77	5.92	6.48	6.87
9	3.20	3.95	4.41	4.76	5.02	5.24	5.43	5.59	5.74	6.28	6.64
10	3.15	3.83	4.33	4.65	4.94	5.12	5.30	5.46	5.60	6.11	6.47
11	3.11	3.82	4.26	4.57	4.82	5.03	5.20	5.35	5.49	5.98	6.33
12	3.08	3.77	4.20	4.51	4.75	4.95	5.12	5.27	5.39	5.88	6.21
13	3.06	3.73	4.15	4.45	4.69	4.88	5.05	5.19	5.32	5.79	6.11
14	3.03	3.70	4.11	4.41	4.64	4.83	4.99	5.13	5.25	5.71	6.03
15	3.01	3.67	4.08	4.37	4.59	4.78	4.94	5.08	5.20	5.65	5.96
16	3.00	3.65	4.05	4.33	4.56	4.74	4.90	5.03	5.15	5.59	5.90
17	2.98	3.63	4.02	4.30	4.52	4.70	4.86	4.98	5.11	5.54	5.84
18	2.97	3.61	4.00	4.28	4.49	4.67	4.82	4.96	5.07	5.50	5.79
19	2.96	3.59	3.98	4.25	4.47	4.65	4.79	4.92	5.04	5.46	5.75
20	2.95	3.58	3.96	4.23	4.45	4.62	4.77	4.90	5.01	5.43	5.71
24	2.92	3.53	3.90	4.17	4.37	5.54	4.68	4.81	4.92	5.32	5.59
30	2.89	3.49	3.85	4.10	4.30	4.46	4.60	4.72	4.82	5.21	5.47
40	2.86	3.44	3.79	4.04	4.23	4.39	4.52	4.63	4.73	5.11	5.36
60	2.83	3.40	3.74	3.98	4.16	4.31	4.44	4.55	4.65	5.00	5.24
120	2.80	3.36	3.68	3.92	4.10	4.24	4.36	4.47	4.56	4.90	5.13
∞	2.77	3.31	3.63	3.86	4.03	4.17	4.29	4.39	4.47	4.80	5.01

Table III.3.3 $q(a=0.01)$

ϕ \ m	2	3	4	5	6	7	8	9	10	15	20
1	90.00	135.00	164.00	186.00	202.00	216.00	227.00	237.00	246.00	227.00	298.00
2	14.00	19.00	22.30	24.70	26.60	28.20	29.50	30.70	31.70	35.40	37.90
3	8.26	10.60	12.20	13.30	14.20	15.00	15.60	16.20	16.70	18.50	19.80
4	6.51	8.12	9.17	9.96	10.60	11.10	11.50	11.90	12.30	13.50	14.40
5	5.70	6.97	7.80	8.42	8.91	9.32	9.67	9.97	10.2	11.2	11.9
6	5.24	6.33	7.03	7.56	7.97	8.32	8.61	8.87	9.10	9.95	10.5
7	4.95	5.92	6.54	7.01	7.37	7.68	7.94	8.17	8.37	9.12	9.65
8	4.74	5.63	6.20	6.63	6.96	7.24	7.47	7.68	7.87	8.55	9.03
9	4.60	5.43	5.96	6.35	6.66	6.91	7.13	7.32	7.49	8.13	8.57
10	4.48	5.27	5.77	6.14	6.48	6.67	6.87	7.05	7.21	7.81	8.22
11	4.39	5.14	5.62	5.97	6.25	6.48	6.67	6.84	6.99	7.56	7.95
12	4.32	5.04	5.50	5.84	6.10	6.32	6.51	6.67	6.81	7.36	7.73
13	4.26	4.96	5.40	5.73	5.98	6.19	6.37	6.53	6.67	7.19	7.55
14	4.21	4.89	5.32	5.63	5.88	6.08	6.26	6.41	6.54	7.05	7.39
15	4.17	4.83	5.25	5.56	5.80	5.99	6.16	6.31	6.44	6.93	7.26
16	4.13	4.78	5.19	5.49	5.72	5.92	6.08	6.22	6.35	6.82	7.15
17	4.10	4.74	5.14	5.43	5.66	5.85	6.01	6.15	6.27	6.73	7.05
18	4.07	4.70	5.09	5.38	5.60	5.79	5.94	6.08	6.20	6.65	6.96
19	4.05	4.67	5.05	5.33	5.55	5.73	5.89	6.02	6.14	6.58	6.89
20	4.02	4.64	5.02	5.29	5.51	5.69	5.84	5.97	6.09	6.52	6.82
24	3.96	4.54	4.91	5.17	5.37	5.54	5.69	5.81	5.92	6.33	6.61
30	3.89	4.45	4.80	5.05	5.24	5.40	5.54	5.65	5.76	6.14	6.41
40	3.82	4.37	4.70	4.93	5.11	5.27	5.39	5.50	5.60	5.96	6.21
60	3.76	4.28	4.60	4.82	4.99	5.13	5.25	5.36	5.45	5.79	6.02
120	3.70	4.20	4.50	4.71	4.87	5.01	5.12	5.21	5.30	5.61	5.83
∞	3.64	4.12	4.40	4.60	4.76	4.88	4.99	5.08	5.16	5.45	5.65

Table III.4 $d(n, l)$ and $\phi(n, l)$

n		1	2	3	4	5	10	15	20	25	30	l > 5	l = ∞
2	ϕ	1.0	1.9	2.8	3.7	4.6	9.0	13.4	17.8	22.2	26.5	$0.876l + 0.25$	1.128
	d	1.41	1.28	1.23	1.21	1.19	1.16	1.15	1.14	1.14	1.14	$1.128 + 0.32/l$	
3	ϕ	2.0	3.8	5.7	7.5	9.3	18.4	27.5	36.6	45.6	34.7	$1.815l + 0.25$	1.693
	d	1.91	1.81	1.77	1.75	1.74	1.72	1.71	1.70	1.70	1.70	$1.693 + 0.23/l$	
4	ϕ	2.9	5.7	8.4	11.2	13.9	27.6	41.3	55.0	68.7	82.4	$2.738l + 0.25$	2.059
	d	2.24	2.15	2.12	2.11	2.10	2.08	2.07	2.06	2.06	2.06	$2.059 + 0.19/l$	
5	ϕ	3.8	7.5	11.1	14.7	18.4	36.5	54.6	72.7	90.8	108.9	$3.623l + 0.25$	2.326
	d	2.48	2.40	2.38	2.37	2.36	2.34	2.33	2.33	2.33	2.33	$2.326 + 0.16/l$	
6	ϕ	4.7	9.2	13.6	18.1	22.6	44.9	67.2	89.6	111.69	134.2	$4.466l + 0.25$	2.534
	d	2.67	2.60	2.58	2.57	2.56	2.55	2.54	2.54	2.54	2.54	$2.534 + 0.14/l$	
7	ϕ	5.5	10.8	16.0	21.3	26.6	52.9	79.3	105.6	131.9	158.3	$5.267l + 0.25$	2.704
	d	2.83	2.77	2.75	2.74	2.73	2.72	2.71	2.71	2.71	2.71	$2.704 + 0.13/l$	
8	ϕ	6.3	12.3	18.3	24.4	30.4	60.6	90.7	120.9	151.0	181.2	$6.031l + 0.25$	2.847
	d	2.96	2.91	2.89	2.88	2.87	2.86	2.85	2.85	2.85	2.85	$2.847 + 0.12/l$	
9	ϕ	7.0	13.8	20.5	27.3	34.0	67.8	101.6	135.3	169.2	203.0	$6.759l + 0.25$	2.970
	d	3.0	3.02	3.01	3.00	2.99	2.98	2.98	2.98	2.97	2.97	$2.970 + 0.11/l$	
10	ϕ	7.7	15.1	22.6	30.1	37.5	74.3	112.0	149.3	186.6	223.8	$7.453l + 0.25$	3.078
	d	3.18	3.13	3.11	3.10	3.10	3.09	3.08	3.08	3.08	3.08	$3.078 + 0.10/l$	

Table III.5

$n-2$	5%	1%	$n-2$	5%	1%	$n-2$	5%	1%
1	0.997	1.000	16	0.468	0.590	35	0.325	0.418
2	0.950	0.990	17	0.456	0.575	40	0.304	0.393
3	0.878	0.959	18	0.444	0.561	45	0.288	0.372
4	0.811	0.917	19	0.433	0.549	50	0.273	0.354
5	0.754	0.874	20	0.423	0.537	60	0.250	0.325
6	0.707	0.834	21	0.413	0.526	70	0.232	0.302
7	0.666	0.798	22	0.404	0.515	80	0.217	0.283
8	0.632	0.765	23	0.396	0.505	90	0.205	0.267
9	0.602	0.735	24	0.383	0.496	100	0.195	0.254
10	0.576	0.708	25	0.381	0.487	125	0.174	0.228
11	0.553	0.684	26	0.374	0.478	150	0.159	0.208
12	0.532	0.661	27	0.367	0.470	200	0.138	0.181
13	0.514	0.641	28	0.361	0.463	300	0.113	0.148
14	0.497	0.623	29	0.355	0.456	400	0.098	0.128
15	0.482	0.606	30	0.349	0.449	1 000	0.062	0.081

Reference

[1] Statistics group of Institute of mathematics, Chinese Academy of Sciences, ANOVA, Science Press, 1977.

[2] Statistical group of Institute of mathematics, Chinese Academy of Sciences, commonly used mathematical statistics method, Science Press, 1973.

[3] Experimental design group of mathematics department, Peking University, "orthogonal experiment method", science popularization press, 1978.

[4] Zhu Weiyong, application of orthogonal and regressive orthogonal experimental method, Liaoning people's publishing house, 1978.

[5] Sun Changming. Application of orthogonal experiment in agricultural science experiment. Agricultural Publishing House, 1978.

[6] 《Concrete and Reinforced Concrete in hot Countries》, 'Proceedings of international RILEM Symposium', 1971, Vol. 1.

[7] Robert. W. Previte, 《Concrete Slump loss》"ACI journal", August 1977.

[8] ACI Committee 207: 《Symposium on mass Concrete》, ACI publication 1963.

[9] Building materials science research and Research Institute, concrete water reducing agent, China Construction Industry Press, 1979.

[10] Zhou Huazhang, mathematical statistics of industrial technology applications, people's education press.

[11] Compilation of this book, "orthogonal experimental design method", Shanghai science and Technology Press, 1979.3.

[12] Zhang Tongbiao, Tian Kewen, Zhang Qing, "the influence of mineral composition of clinker cement on the strength of cement", 1963.4.